Heritage Education for Climate Action

Research in Architectural Education Set
coordinated by
Lazaros E. Mavromatidis

Volume 2

Heritage Education for Climate Action

Edited by

Irene G. Curulli
Deniz Ikiz Kaya
Arghavan Khaefi

iSTE

WILEY

First published 2023 in Great Britain and the United States by ISTE Ltd and John Wiley & Sons, Inc.

ISTE Ltd
27-37 St George's Road
London SW19 4EU
UK

www.iste.co.uk

John Wiley & Sons, Inc.
111 River Street
Hoboken, NJ 07030
USA

www.wiley.com

Library of Congress Control Number: 2023943143

British Library Cataloguing-in-Publication Data
A CIP record for this book is available from the British Library
ISBN 978-1-78630-903-7

Contents

Introduction

I.1. Heritage education and climate change

Focusing on heritage education and training for climate action, this book discusses the role that heritage education plays regarding climate issues. It does this with a wide set of teaching and learning examples, methods and tools on heritage and resilience building, vulnerability and risk assessment, climate change adaptation, mitigation and policymaking.

"Education is crucial to promote climate action. It helps people understand and address the impacts of the climate crisis, empowering them with the knowledge, skills, values, and attitudes needed to act as agents of change" (UNESCO-Climate Change Education 2023). To address this goal, many environmental education programs have developed over the past years, engaging learners through inquiry and constructive learning and offering training and resources to promote climate knowledge and climate awareness.

However, the development of climate adaptation competencies requires a foundation of knowledge from a number of scientific areas and from different perspectives, as well as understanding and being able to bridge and apply system thinking to a range of issues. In short, the acquisition of climate adaptation competencies requires cross-disciplinary knowledge which will enhance students' creative thinking and empower them to come up with their own ideas, all based on an integrated disciplinary learning.

Introduction written by Irene G. CURULLI, Deniz IKIZ KAYA and Arghavan KHAEFI.

Heritage education can play an important role in developing climate adaptation competencies so as to foster positive climate change dialogs, shift attitudes and inspire action.

According to The Council of Europe, heritage education is an educational practice in which culture and heritage function as teaching and learning resources (Council of Europe 2005). Thus, heritage education provides life-enhancing learning experiences and inspiration as well as opportunities to develop our sense of well-being and belonging (European Commission 2019). Therefore, it is essential to make use of the power of heritage to create public engagement/education on climate change by developing new critical pedagogical methods in education (ICOMOS-Climate Change and Cultural Heritage Working Group International 2019). Education that addresses the joint issue of climate change and heritage conservation contributes to the acknowledgment of heritage values. It also helps us understand not only the climate risks to cultural heritage but also the positive role it can play as a source of resilience for communities, as well as an invaluable repository of information and strategies to address climate change (ICOMOS-Climate Change and Cultural Heritage Working Group International 2019).

Cultural heritage is increasingly acknowledged to be something that helps shift mindsets and open up new practices and policy-oriented horizons to drive climate resilient development. However, their integration into climate science and education has so far received limited attention. Given diverse local circumstances and priorities, existing education, training and capacity-building efforts that bridge heritage and climate change have been fragmented, project-oriented and works-in-progress to address varied emerging needs and challenges. There is thus an increasing need and demand to adapt heritage education and training so that they address climate issues in an interdisciplinary manner.

The 2015 Paris Agreement, which recognizes adaptation as a global challenge and goal to be achieved across sectors through national and international cooperation, highlighted the importance of capacity building. Nonetheless, it is challenging to accommodate teaching and learning activities that are multi- and transdisciplinary, as well as inclusive and holistic, into traditional higher education curricula and models.

Furthermore, in a recent report that discusses the role of cultural heritage in the European Green Deal and climate resilience, the European Commission acknowledges the significance of bridging the research and education on cultural heritage and climate change, and yet, the heritage-climate change link is barely included in the existing educational structures (Council of Europe 2022). This is why the Commission puts an emphasis on systematic educational and training schemes that foster interdisciplinary and dynamic teaching and learning practices.

I.2. Taking action

Smith (2006), Harvey (2001) and Byrne (2014) interpret heritage as a process or series of practices that connect the present to the past, often favoring processes of change (Holtorf 2014; Dawdy 2016). In his critical assessment of heritage ontologies, Harrison (2018) addresses heritage as an interdisciplinary domain that takes into account multiple denotations and relevant practices that produce "multiple modes of being […] that work towards the production of multiple specific futures". How much this disciplinary and ontological multiplicity is reflected in the curricula for heritage education remains to be said, however. The future of the field demands a more dynamic and adaptive approach toward heritage.

In their critical discussion over the heritage-climate change nexus, Harvey and Perry (2015) call upon a new heritage approach that crosses the existing ontological boundaries that is flexible and comprehensive regarding the notions of change embracing uncertainties, loss and damage in heritage, and is creative enough to propose adaptive solutions. In this wider context, culture and heritage can act as catalysts for positive transformational change by connecting people to places, encouraging a sense of place and belonging, and fostering social inclusion while accommodating integrative knowledge systems and ontological pluralism. Thus, heritage can serve as a powerful resource for deepening our collective knowledge, paving the way for new creative solutions to tackle the impact of climate change and contributing to transformational pathways.

While the role that cultural heritage plays in this transition for climate action has been newly recognized in national and supranational policymaking, there is still a gap that exists between policy, practice and local capacity building for heritage-oriented climate change mitigation and adaptation. There is thus an increasing need and demand to adapt heritage education and training to address climate issues in an interdisciplinary manner.

So, how can we advance heritage education in a way so that it can contribute to sustainable climate action?

To explore, develop and discuss new and innovative education perspectives on the above-mentioned heritage challenges, researchers from six European universities organized the webinar "Boosting resilience of Europe's cultural heritage to climate change" in May 2022. The webinar was part of the Erasmus+ KA2 project entitled "Education for Climate Resilient Education Heritage & Architecture (e-CREHA)", which aimed to develop and test innovative pedagogical approaches and teaching/learning methods that integrate heritage and climate science. The call for papers invited researchers, tutors, students, experts, professionals and policymakers to present best practices on cross-cultural pedagogies in Heritage Learning focusing on the vulnerability, impact, mitigation and adaptation of built heritage.

This book provides a rich collection of detailed case studies that presents a broad scope of pedagogical strategies that rely on theory and practice. The book covers a wide array of knowledge areas, international experiences, cases and practices. Stemming from the ontological and disciplinary plurality in the heritage-climate change nexus, it thus brings together lessons and experiences from multiple disciplines ranging from architecture, engineering, heritage studies, geography, tourism and social sciences in a cross-disciplinary manner. Bringing together knowledge, practices and experiences from different disciplines, this book fills in the gap in the heritage field that intends to adapt and transition heritage education toward sustainable development and climate action.

I.3. Roadmap

This book consists of three parts. Part 1 examines heritage education, training and capacity building pedagogies and approaches toward climate awareness, adaptation and mitigation.

In Part 1, Chapter 1 by Pelin Gürol-Öngören, Sibel Acar, Asli Özbek and Burçin Yilmaz, begins with an inventory and analysis of higher education courses offered by European universities in the field of built environment, landscape and heritage studies. The authors reflect upon how the existing courses cover the topics of resilience, impact, vulnerability and adaptation regarding the interconnection of built heritage and climate change, and point out the limitations in the existing curriculums.

Chapter 2, by Lazaros Mavromatidis, opens the debate for transdisciplinary knowledge and provides an example of their teaching that incorporates thermodynamics theories at the primary stages of architectural design in order to create a more comprehensive understanding of the "spatial hypotheses" that can lead to designs that are more sustainable and better suitable to the environment. Chapter 3, by Anna Kaiser, exemplifies multidisciplinary collaboration in training where multiple professional groups work together to define strategies for cultural and natural heritage protection that is at risk. Chapter 4, by Yavor Dankov, Yakim Deykov and Andjela Dankova, brings into discussion the importance of multiscalar education that targets different age groups, ranging from primary school education to higher education, in order to extend awareness and accessibility on heritage and climate education.

Chapter 5, by Boyan Bontchev, Valentina Terzieva, Albena Antonova, Yavor Dankov and Dessislava Vassileva, presents educational video games as a pedagogical tool to raise awareness and reach a broader audience to engage them in heritage and climate action through culture and entertainment. It also exemplifies how different disciplines, such as computer sciences, can also contribute to heritage and climate education.

The JTHECA experience described by Susan Ross in Chapter 6 further highlights the importance of building synergy among multiple disciplines, knowledge areas and pedagogical approaches that also take into account indigenous knowledge and experiences regarding equity, inclusiveness and just transition. The last chapter in Part 1, Chapter 7 by Irene G. Curulli, Deniz Ikiz Kaya, Rossella Nocera and Luciano De Bonis, discusses a multidisciplinary blended learning methodology for climate resilient heritage by addressing the design, programming, supervision and evaluation processes of the e-learning course, the e-CREHA Erasmus+ KA203 program, which this book volume is inspired by.

Part 2 presents examples and teaching practices that focus on cultural heritage vulnerability, risk assessments and the impact of climate change. Chapter 8, by Luciano De Bonis, Rossella Nocera and Giovanni Ottaviano, concentrates on territorial heritage and emphasizes reactivating the co-evolutionary dynamic of mutual adaptation between human beings and non-human environmental factors, in order to maintain a resilient heritage landscape. Focusing also on cultural landscapes at risk, Maria Gabriella Travato in Chapter 9 demonstrates student work on threatened landscapes in Lebanon as a case study. Another case study of cultural landscape is provided by Eva Stark and Lina Bekker in Chapter 10, exploring the transformation of watermill heritage landscapes in the Netherlands.

Part 3 focuses on heritage and climate adaptation practice and presents a wide array of methods, tools and policy approaches that are designed, evaluated and implemented. In Chapter 11, Deniz Ikiz Kaya and Irene G. Curulli demonstrate the use of real-life cases and collaborations as a pedagogical approach to investigating the climate adaptation of historic watermills in the Netherlands. In Chapter 12, Nihan Bulut calls attention to 20th-century architecture and the need to create a framework of intervention strategies for better climate adaptation and discusses the role that education can play.

Paloma Guzman and Jelle Buijs, in Chapter 13, provide a comprehensive and critical discussion on transforming existing pedagogies in heritage education through the integration of a value-based approach to foster knowledge integration and more critical, reflexive and problem-based pedagogies. In Chapter 14, Scott Bremer exemplifies this by integrating seasonal cultures into education and awareness-raising initiatives. Finally, in Chapter 15, Maya Ishizawa Escudero, Eugene Jo and Rohit Jigyasu describe the capacity-building efforts and training programs offered by international heritage organizations that incorporate climate science and knowledge into policies and capacity building for World Heritage sites, such as ICCROM, UNESCO, ICOMOS and IUCN.

The chapters are followed by reflections on the current national education programs and agendas in several European countries, including France (Appendix 1 by Lazaros Mavromatidis), the Netherlands (Appendix 2 by Irene G. Curulli and Deniz Ikiz Kaya), Italy (Appendix 3 by Rossella Nocera, Luciano De Bonis and Giovanni Parisani) and Turkey (Appendix 4 by Pelin Gürol-Öngören, Sibel Acar, Aslı Özbek, Burçin Yilmaz and Irem Küçük).

This volume brings together a variety of pedagogical approaches that promote an inquiry-based learning that encourages students to engage in problem-solving and experiential learning through exploration and high-level questioning concerning the nexus of heritage and climate change. Furthermore, these pedagogies foster collaboration and partnerships between experts, institutions and communities to leverage resources and expertise to promote sustainable climate action through heritage education. Thus, this book is a source of suggestions and ideas for scholars, educators and professionals who want to develop future climate leadership.

I.4. References

Byrne, D. (2014). *Counter Heritage: Critical Perspectives on Heritage Conservation in Asia.* Routledge, Abingdon.

Council of Europe (2005). Convention on the value of cultural heritage for society [Online]. Available at: https://www.coe.int/en/web/culture-and-heritage/faro-convention.

Council of Europe (2022). The Faro Convention's role in a changing society: Building on a decade of advancement. Document, Council of Europe.

Dawdy, S.L. (2016). *Patina: A Profane Archaeology*. University of Chicago Press.

European Commission (2019). Cultural heritage and education [Online]. Available at: https://culture.ec.europa.eu/cultural-heritage/cultural-heritage-in-eu-policies/cultural-heritage-and-education [Accessed May 25, 2023].

European Commission Directorate-General for Education, Youth, Sport and Culture (2019). European framework for action on cultural heritage [Online]. Available at: https://data.europa.eu/doi/10.2766/949707.

Harrison, R. (2018). On heritage ontologies: Rethinking the material worlds of heritage. *Anthropological Quarterly*, 91(4), 1365–1383.

Harvey, D.C. (2001). Heritage Pasts and Heritage Presents: Temporality, meaning and the scope of heritage studies. *International Journal of Heritage Studies*, 7(4), 319–338.

Harvey, D.C. and Perry, J. (eds) (2015). *The Future of Heritage as Climates Change Loss, Adaptation and Creativity.* Routledge, Abingdon.

Holtorf, C. (2014). Averting loss aversion in cultural heritage. *International Journal of Heritage Studies*, 21(4), 405–421 [Online]. Available at: https://doi.org/10.1080/13527258.2014.938766.

ICOMOS-Climate Change and Cultural Heritage Working Group International (2019). The future of our pasts: Engaging cultural heritage in climate action. Technical report, International Council on Monuments and Sites, Paris [Online]. Available at: https://openarchive.icomos.org/id/eprint/2459/.

Smith, L. (2006). *Uses of Heritage*. Routledge, New York.

UNESCO-Climate Change Education (2023). Climate change education [Online]. Available at: https://www.unesco.org/en/education-sustainable-development/climate-change [Accessed 8 May, 2023].

PART 1

Heritage Education Pedagogies and Approaches for Climate Action

1

Cross-Disciplinary Pedagogies: Education on Heritage and Climate Change in Europe

1.1. Introduction

Cultural heritage in Europe, as well as worldwide, is threatened by climate change, which impairs its integrity and threatens its values. In addition to amplifying existing vulnerabilities and threats, climate change creates new risks to cultural heritage (Masson-Delmotte et al. 2021). Cultural heritage includes not only physical artifacts, but also the knowledge and skills gained through human practices, representations and experiences related to the objects and spaces recognized by communities as cultural heritage (UNESCO n.d.). Cultural heritage assets such as historical buildings, archeological sites and monuments, as well as their contents and collections, are legacy items from our past that lend a sense of place, identity and esthetic well-being to local communities (Sesana et al. 2021). Accordingly, the value of cultural heritage extends beyond its physical presence. People connect with places through their cultural heritage. It is a part of people's identities and inherent in almost every aspect of their life. As a result, it fosters a sense of belonging, promotes social inclusion and brings individuals together to collaborate more effectively. Thus, cultural heritage can be used to raise awareness about climate change and the need for climate action. Heritage may be impacted profoundly by climate change. However, it can also be a source of creativity, innovation and knowledge that can inspire and guide transitions to low-carbon, climate-resilient futures (Potts 2021, p. 6). This dual approach to climate change is promoted by the International Council

---

Chapter written by Pelin GÜROL-ÖNGÖREN, Sibel ACAR, Asli ÖZBEK and Burçin YILMAZ.
Based on the report entitled "Cross-Disciplinary Pedagogies of the e-CREHA project" submitted in November 2021. Please see the website for more info: https://www.ecreha.org.

on Monuments and Sites (ICOMOS), which seeks to both address the risks climate change poses to cultural heritage, and promote heritage as a source of resilience and a resource for climate action (ICOMOS n.d.).

The role of architecture in reducing climate change effects (transition to zero-emission, energy-efficient and resilient buildings) is crucial. Over the past decade, climate change issues have become increasingly relevant to architectural design, urban planning, heritage studies and relevant disciplines. The Architects' Council of Europe (ACE) has urged architects to become more prominent in climate adaptation. ACE promotes a holistic approach to the built environment that serves all stakeholders economically, socially, environmentally and culturally (ACE 2021). Knowledge, expertise and experience from different fields and disciplines must be integrated when addressing climate change and embracing heritage as a resource and driver for sustainable development. This situation requires a focus on heritage and climate change in architectural education. This study focuses on the built environment, defining human-made environments, which also indicates cultural heritage that includes monuments, groups of buildings, sites and cultural landscapes, which are "combined works of nature and of man" (UNESCO 2011).

1.2. Climate change and heritage education

Climate change effects are well known worldwide, but research and quantification of their impacts on cultural heritage are still limited. To manage cultural heritage at risk, it is essential to implement effective strategies at all levels, from government policies to field practice. This requires awareness of climate change action, skilled human resources and capacity building. Professionals with solid skills in sustainable design and resilience are needed to develop and apply mitigation and adaptation measures. Therefore, it is critical to apply, update and develop professionals' skills continuously. In this regard, education and climate change are becoming more relevant and intertwined. Climate change's effects on heritage must be an integral part of architectural education. The curricula in architecture schools must emphasize the built heritage's vulnerability to climate change and the heritage's role in developing mitigation and adaptation strategies.

There is a complex relationship between climate change and cultural heritage (science, economics, society, politics, morals and ethics). This research shows a disparity between research, education and practice. To overcome this disparity, a multidisciplinary and holistic approach is needed in research, education and practice. Additionally, it is necessary to use innovative learning methods such as information and communication technology (ICT)-based learning processes in heritage and climate change education.

The Education for Climate-Resilient European Heritage Architecture (e-CREHA) project addresses the gaps. It aims to constitute the knowledge between education and research recommended by the 2018 European Union (EU) modernization agenda. The project's objective is to foster cooperation between research and education, theory and professional practice to enhance our understanding of how cultural heritage and design can contribute to climate change. The project aims to train and educate the next generation of tutors, researchers and professionals to help create a climate-resilient European cultural heritage. Multi-disciplinary experts and students from different European Higher Education Institutions (HEIs) have collaborated to formulate an innovative e-learning course and methodology focusing on developing climate resilience for European cultural heritage. The course focuses on resilient architectural heritage and climate change, and adopts an innovative approach and new interpretations of the content and methods. It is designed for graduate students in built environment-related programs to expand their intellectual resources and provide them with a new set of skills. The course includes modules with lectures, quizzes/games, video clips, assignments and students' virtual forum(s). Four main concepts are covered in the course:

RESILIENCE (built heritage and resilient concepts): The term "resilience" is defined by the Intergovernmental Panel on Climate Change (Mach et al. 2014, p. 127):

> The capacity of social, economic, and environmental systems to cope with a hazardous event or trend or disturbance, responding or reorganizing in ways that maintain their essential function, identity, and structure, while also maintaining the capacity for adaptation, learning, and transformation.

IMPACT (built heritage and impact of climate change): The IPCC Special Report on Climate Change and Land (van Diemen 2019, p. 815) defines the term "impact":

> The consequences of realized risks on natural and human systems, where risks result from the interactions of climate-related hazards (including extreme weather and climate events), exposure, and vulnerability. Impacts generally refer to effects on lives; livelihoods; health and well-being; ecosystems and species; economic, social, and cultural assets; services (including ecosystem services); and infrastructure. Impacts may be referred to as consequences or outcomes and can be adverse or beneficial.

VULNERABILITY (heritage as a resource for climate change): The IPCC SRCCL (van Diemen 2019, p. 826) explains "vulnerability":

> The propensity or predisposition to be adversely affected. Vulnerability encompasses a variety of concepts and elements, including sensitivity or susceptibility to harm and lack of capacity to cope and adapt.

ADAPTATION (resilient heritage solutions): The IPCC SRCCL (van Diemen 2019, p. 804) defines "adaptation":

> In *human systems*, the process of adjustment to actual or expected *climate* and its effects, in order to moderate harm or exploit beneficial opportunities. In natural systems, the process of adjustment to actual climate and its effects; human intervention may facilitate adjustment to expected climate and its effects.

The report "Cross-disciplinary Pedagogies" (CDP), as the first intellectual output of the e-CREHA project, attempts to understand the current state of the art on how heritage in architectural education is responding to climate change. This research relies on the research and analysis of current interdisciplinary learning/teaching methods and approaches in different categories related to cultural heritage and climate change.

1.3. Research and analysis

Implementation of the e-CREHA e-learning course is based on a blended learning approach and setting up a new/experimental course. CDP aimed to specify and assess current approaches, interpretations and mediums in cross-disciplinary studies addressing heritage and climate change. In that framework, an Excel form was created for partners to list relevant courses and collect data on their content and methods. All partners filled in this table[1] by searching architecture/built environment schools and related academic institutions in their network. They also used search engines (Google, Bing and Baidu), Web of Science, Scopus and online learning platforms to search the Massive Open Online Course (MOOC) between February and August 2021.

---

1 The research includes a list of current courses, games, documentaries and, in addition, climate governance, and the role of plans and policies. This chapter summarizes the results for courses. For the results of the other categories, please see the O1 CDP report (https://www.ecreha.org/research).

A convenient sampling based on geographic specialization was applied, including the methods/tools adopted in consortium countries, starting from their home countries which are most accessible to the partners. In the data collection phase, the limitations were the language barrier, the inaccessibility of related information, lack of information, or out-of-date online data about educational programs on the websites. For the qualitative analysis, in some cases, the nature of the data was not descriptive in terms of the contents and methods that they used[2].

The research results include 37 courses offered by institutions in nine different countries[3]. As a first analysis, we looked at course distribution by country, institution and year. It reveals that there is a greater interest/awareness/study of climate change and built heritage in The Netherlands and France than in other partner countries (Figure 1.1).

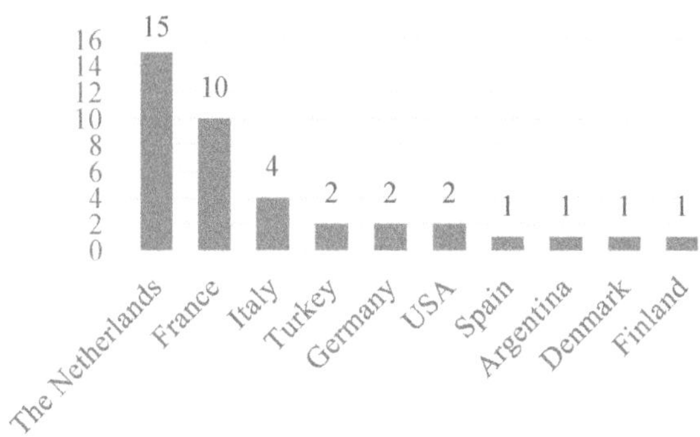

Figure 1.1. *Distribution of courses by country*

There has been an increase in the number of courses since 2017 (Figure 1.2). The reason may be that the impact of climate change has become more severe and threatening over the years. An increase in the number of courses coincides with the increasing international policies on climate change. In line with the Paris Agreement

2 For more info about the research (data, tables, graphs and more), please see the O1 CDP report (https://www.ecreha.org/research).

3 In addition to these courses, we reached two master programs, "Climate-adapted Construction and Operation" at BTU Cottbus (Germany) and "Radical Sustainable Architecture" at Aarhus School of Architecture (Denmark), and three other design courses: "Nationaal Centrum Duurzaamheid", "De Eeuw van de Amateur-Wildcard Studio-focus on vernacular architecture" and "Weerstudio" by Academie Rotterdam (RAvB, The Netherlands). They are not included in the content analysis presented in Table 1.1, since we did not have sufficient information about their contents.

and the Sustainable Development Goals (SDGs), Europe has committed to cutting greenhouse gas emissions and transforming its economy and society in a zero-emission direction. Consequently, capacity-building efforts have made the built environment and environmental systems more resilient. Academies and institutes have begun to develop courses addressing issues associated with climate change.

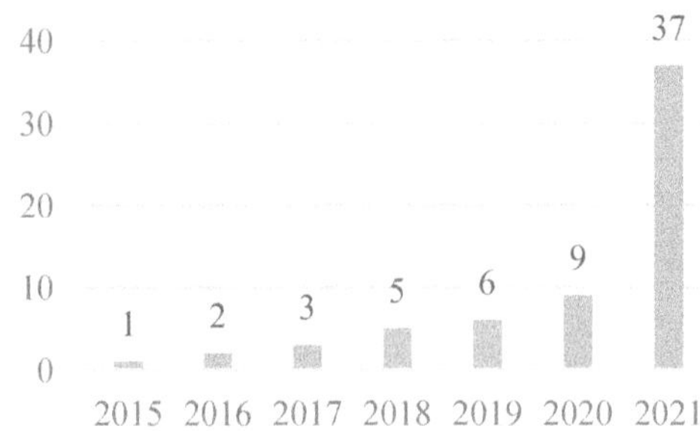

Figure 1.2. *Number of courses by year*

As part of our study, we also examined course types, contents and target groups, as well as learning environments, methods and tutors' expertise areas. The courses vary in terms of their scopes, objectives, target groups, concepts, duration, pedagogical methods and learning environments. CDP shows that most courses are designed by HEIs (Figure 1.3). Regarding their target groups, the courses can roughly be classified as summer schools, MOOCs and undergraduate and graduate courses. Summer Schools are short-term events aimed at a broader audience than students (Figure 1.4). Regarding target groups, climate change and built heritage are generally considered graduate-level topics.

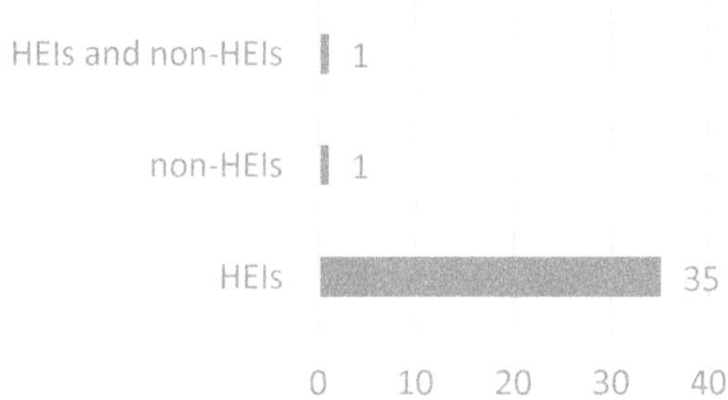

Figure 1.3. *Number of courses offered by HEIs and non-HEIs*

Figure 1.4. *Course type by number*

Regarding the learning environment of the courses, there are lecture-based and studio courses. Most studio courses focus on case studies, including fieldwork and workshops to develop adaptation and mitigation strategies. Courses focusing on heritage impact assessment methods, heritage vulnerability and conservation mostly offer a model of lectures and seminars together. Most of the courses in the curriculum are instructed by architects and urban planners. These courses have limited cooperation with other fields such as engineering, sociology, history, economy, management, environmental and climate sciences. MOOCs, on the other hand, are often taught by groups of experts with backgrounds in architecture, urban planning, civil engineering, history, environmental economics, social sciences and climate change.

When we analyzed courses according to their focus and extent of dealing with climate change and heritage, two main categories are evident: the courses offered in the built environment-related programs (A) and the climate science courses (B) (Table 1.1).

Heritage and climate change-related courses offered in architecture/built environment/urban planning departments present varieties regarding their scopes and the extent of their studies. Four categories of courses are observed in this category. A1 courses are the built environment courses. Neither climate change nor heritage is the main focus, yet climate change's impact on heritage is one of the issues discussed. A2 courses deal with the issues of the built environment and climate change. However, they do not devote the entire course to heritage and climate change studies. These courses are offered in Architecture, Civil Engineering, Building and Energy Technology departments at both the undergraduate and graduate levels. While the broader focus of these courses is on climate change and the built environment, climate-resilient heritage is a small part of them.

Courses on built heritage constitute the third category (A3). These courses focus on built heritage throughout the entire semester/trimester/quarter. One of the topics they discuss is climate-resilient heritage as part of a broader discussion. Most of these courses are offered in architecture and urban planning graduate programs provided by architects and urban planners. Sometimes, engineers, landscape architects, historians and experts from different areas join the team as well. These courses differ in approach and focus; thus, they are divided into two subcategories. There are some design courses (A3.1) that develop adaptation proposals for the heritage at various scales, including rehabilitation and retrofitting. Another category of courses (A3.2) is mainly offered by graduate programs in conservation, preservation or building science. Within the context of conservation, these courses discuss the impact of climate change, vulnerability and adaptation of built heritage.

The last category of courses on built environment-related programs (A4) addresses both climate change and built heritage. The courses focus on climate change and built heritage throughout the semester/quarter/trimester. These courses can be examined under three subcategories (A4.1, A4.2, A4.3). The first subcategory (A4.1) includes courses offered by graduate programs in conservation, preservation or building science. These courses deal with the risk and impact assessment of climate change on built heritage. Design courses constitute another subcategory (A4.2). Courses in this category give students a critical perspective on vulnerability and adaptation through design to adhere to governmental measures on flood areas and the SDGs (Goals 6, 11, 13). The last subcategory (A4.3) contains courses focusing on heritage assessments and rehabilitation, and also examines multidisciplinary aspects of cultural heritage as a resource for mitigation and adaptation.

The courses offered by the Departments of Natural and Social Sciences on climate change constitute category B. Although these courses do not directly address its impacts on built heritage, climate change is addressed as a phenomenon to be understood through the lens of natural sciences. Climate change has complicated causes and implications rooted in historical, social and economic factors. With a foundation in both natural and social sciences, these courses deal with climate change issues with a focus on cultural geography, history and sociology. Courses in Category B can be evaluated in three subcategories according to their main focuses: climate science courses (B1), courses dealing with social and geographic implications of climate change (B2) and courses addressing the built environment as an agent for mitigation or intensifying climate change (B3).

Categories of courses according to their focus and extent of climate change and cultural heritage	
Main focus of the courses	Title of the courses
A. The built environment and climate change courses	
A1. Built environment courses where neither climate change nor heritage is the main focus. The impact of climate change on heritage is one of the issues discussed	
– Structure and materiality of architecture	– Construction and Durability (ENSA Toulouse, France)
A2. Courses with a focus on the built environment and climate change heritage is one of the issues discussed	
Climate resilient built environment	– Climatic Heterotopias (INSA Strasbourg, France) – Climate-adapted Construction and Operation (BTU Cottbus, Germany) – Radical Sustainable Architecture (Aarhus School of Architecture, Denmark) – Climate Change and Design (Eskişehir Osmangazi – University, Türkiye) – (Re)Imagining Port Cities: Understanding Space, Society, and Culture (TU Delft, The Netherlands, MOOC by Edx)
A3. Courses on built heritage Climate change is just one of the issues addressed	
A3.1 Adaptation through transformation at various scales, including rehabilitation and retrofitting	– Heritage and Architecture Design Studio: Research and Architectural Design (TU Delft, The Netherlands) – Architecture, Heritage, and Eco-construction (ENSA Bretagne, France) – Heritage and Old Buildings (ENSAP Lille, France) – Trans-Form (ENSA Normandie, France) – Architecture and Heritage (ENSA Grenoble, France)
A3.2 Vulnerability, conservation, preservation, restoration	– Geophysics and Cultural Heritage (University of Venice Conservation Science and Technology for Cultural Heritage, Italy) – Environmental Impact on Materials Deterioration and Aging (Università di Bologna Laurea Magistrale in Science for the conservation-restoration of cultural heritage, Italy) – Course on Architectural Heritage (INSA Strasbourg, France) – TD de techniques patrimoniales d'intervention durable (ENSA Clermont-Ferrand, France) – Doctrines Intervention on Built Environment (INSA Marseille, France) – Heritage under Threat (Universiteit Leiden Centre for Global Heritage and Development, The Netherlands, MOOC by Coursera) – Atelier Architectural Restoration Project (Politecnico di Torino, Italy) – World Heritage and Site Management (Kadir Has University, Türkiye)

A4. Built heritage and climate change courses	
A4.1 Vulnerability, adaptation, and conservation	– Climate Change and Risk Assessment for Cultural – Heritage (ICOMOS Argentina, MOOC) – Heritage Impact Assessment Transformation Strategies for Climate Change (BTU Cottbus, Germany)
A4.2 Adaptation and transformation	– Master Project on Vulnerability (TU/e, The Netherlands) – Design course: de Haven van Antwerp (RAvB, The Netherlands)
A4.3 Climate resilient heritage Heritage as an asset for mitigation and adaptation	– Graduation project-topic: Resilience and climate change (TU/e, The Netherlands) – Built-up Heritage Over Time (TU/e, The Netherlands)
B. Climate change courses Heritage is minimal or absent in the content	
B1. Climate science courses	– Making Climate Adaptation Happen (Wageningen University, The Netherlands, MOOC)
B2. Courses dealing with social and geographic implications of climate change (the content of the course based on both natural and social sciences)	– Adaptation Governance (University of Groningen, The Netherlands) – Design of Climate Change Mitigation and Adaptation Strategies (Wageningen University, The Netherlands) – Uncertainty and Climate Change. Challenges for a UNESCO World Heritage City (UNED, Spain, Summer School)
B3. Courses focusing on the built environment as an agent for mitigation or intensifying climate change.	– Climate Change, Health and Architecture (Aalto University, Finland, Summer School) – Planning for Climate Change in African Cities (United Cities and Local Governments of Africa, African Local Government Academy, Erasmus University Rotterdam, Institute for Housing and Urban Development, MOOC by Coursera) – Climate-Responsive Planning and Design (Wageningen University, The Netherlands)

Table 1.1. *Courses according to their focus and extent of climate change and cultural heritage*[4]

For the above courses, two different analyses were made. Figure 1.5 illustrates how many times the concepts of resilience, impact, vulnerability and adaptation appear in their content. Figure 1.6 is a graphic presentation of our results that indicates which of the concepts of resilience, impact, vulnerability and adaptation are studied in conjunction with one another[5]. Resilience, impact, vulnerability and adaptation are mostly addressed together. Resilience, vulnerability and adaptation

4 Available at: https://www.ecreha.org/research.

5 Available at: https://www.ecreha.org/research.

are the second most frequently addressed concepts. Adaptation is the most addressed concept. Vulnerability is most often addressed alongside adaptation.

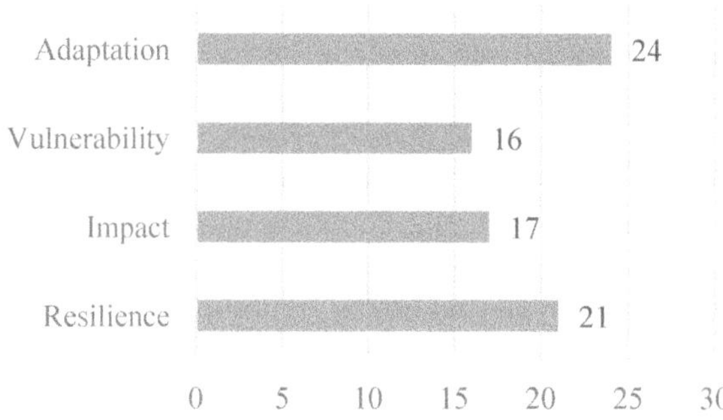

Figure 1.5. *Frequency of each theme*

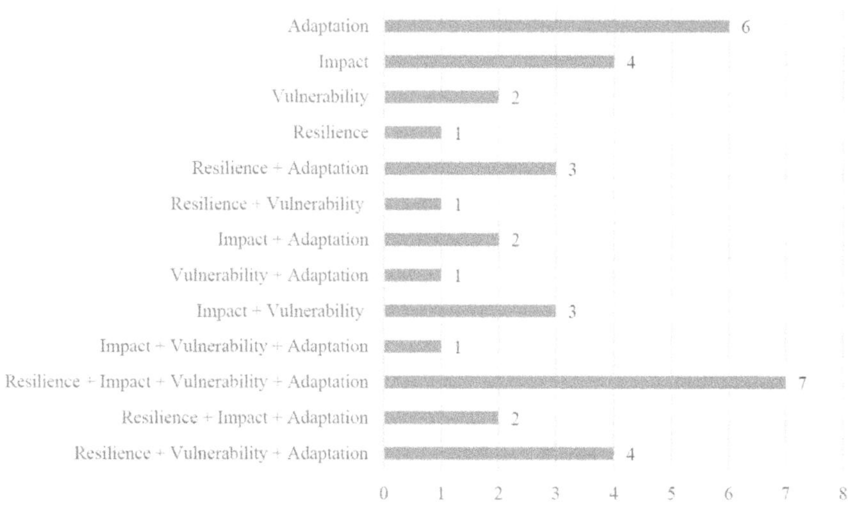

Figure 1.6. *The groups of concepts of resilience, impact, vulnerability and adaptation*

1.4. Conclusion

The research revealed that heritage education addresses climate change issues in several ways with different methods and objectives. While some have a clear focus on mitigation through nature-based solutions, others study in the scope of heritage studies. However, the curricula of HEIs lack a specialized course on the intersection of cultural heritage and climate change. ICOMOS promotes a dual approach to climate change, which addresses the risks climate change poses to cultural heritage by declaring heritage as a resource for climate action and resilience. Yet, this dual approach is rarely considered in the curricula and courses. Understanding the role heritage plays in climate actions and how it contributes to community resilience is important for future architects. In order to take part in climate action, architects need to comprehend how multi-level planning mechanisms and cultural heritage conservation can be integrated. Yet, it can be challenging because of the lack of interdisciplinary knowledge on climate change and cultural heritage in architectural education.

It is obvious that there is a need to bridge the gap between the mechanisms, governance and research and education in order to comprehend multi-level planning mechanisms and deal with potential conflicts between diverse levels of planning frameworks. Future practitioners/policymakers/designers need to be better trained and qualified to apply traditional construction techniques, cultural meanings and implications, and innovative technological approaches in adaptation. Therefore, the curricula should include courses that emphasize how built heritage is vulnerable to climate change and the pivotal role it plays in developing climate-adaptive strategies.

1.5. References

ACE (2021). A sustainable, fair and beautiful built environment to address the climate and biodiversity crisis. Statement presented at "Climate Change & Built Heritage" Conference, ACE, Brussels, October 28, 2021 [Online]. Available at: https://www.ace-cae.eu/fileadmin/user_upload/ACE_CC_CONF_DRAFT_STATEMENT.pdf [Accessed 13 January 2023].

van Diemen, R. (2019). Annex I: Glossary. In *Climate Change and Land: An IPCC Special Report on Climate Change, Desertification, Land Degradation, Sustainable Land Management, Food Security, and Greenhouse Gas Fluxes in Terrestrial Ecosystems*, Shukla, P.R., Skea, J., Calvo Buendia, E., Masson-Delmotte, V., Pörtner, H.-O., Roberts, D.C., Zhai, P., Slade, R., Connors, S., van Diemen, R. et al. (eds). IPCC, Geneva.

ICOMOS (n.d.). ICOMOS work on climate change [Online] Available at: https://www.icomos.org/en/focus/climate-change/60669-icomos-work-on-climate-change [Accessed 10 January 2023].

Mach, K.J., Planton, S., von Stechow, C. (2014). Annex II: Glossary. In *Climate Change 2014: Synthesis Report. Contribution of Working Groups I, II and III to the Fifth Assessment Report of the Intergovernmental Panel on Climate Change*, Pachauri, R.K., Meyer, L.A., Core Writing Team (eds). IPCC, Geneva.

Masson-Delmotte, V., Zhai, P., Pirani, A., Connors, S.L., Péan, C., Berger, S., Caud, N., Chen, Y., Goldfarb, L., Gomis, M.I. et al. (2021). *Climate Change 2021: The Physical Science Basis. Contribution of Working Group I to the Sixth Assessment Report of the Intergovernmental Panel on Climate Change*. Cambridge University Press.

Potts, A. (2021). European Cultural Heritage Green Paper. Document, Europa Nostra, The Hague and Brussels.

Sesana, E., Gagnon, A., Ciantelli, C., Cassar, J., Hughes, J. (2021). Climate change impacts on cultural heritage: A literature review. *WIREs Climate Change*, 12, e710:2.

UNESCO (n.d.). What is intangible cultural heritage? [Online]. Available at: https://ich.unesco.org/en/what-is-intangible-heritage-00003 [Accessed 5 January 2023].

UNESCO (2011). *Preparing World Heritage Nominations*, 2nd edition. UNESCO, Paris.

2

Designing "Climatic Constructal Heterotopias" Using Constructal Thermodynamics: Connecting Theories on Architecture and Thermodynamics

2.1. Introduction and general context

The present paper introduces a novel approach to pedagogy that emphasizes the development of complex urban, social, formalistic, environmental and thermodynamic problems, rather than focusing on the creation of a simple mono-dimensional architectural project. This method involves starting from an apparently paradoxical concept and gradually progressing through a thermodynamic description to make crucial architectural and urban decisions. By conceiving "*climatic constructal heterotopias*", (Figure 2.1) the approach prioritizes research and investigation, rather than fixating on achieved results (Mavromatidis 2019, 2023). This method thus differs from the contemporary approach, where architectural and environmental arguments are often treated as static, final and determinate abstract proposals that hinder progress.

In this proposal, I suggest that the architectural problem should be directly linked to thermodynamics at an early design stage. Thermodynamics, which is the "*science of evolution*", can play a critical role in shaping an architectural project, which is essentially an ever-evolving entity that transforms abstract ideas into concrete spatial objects. The functioning of contemporary thermodynamics relies on hypotheses and measurements made at both the microscopic level of individual particles and the macroscopic level of "*large*" systems (Baker 1997). Similarly, architecture is born

---

Chapter written by Lazaros MAVROMATIDIS.

out of spatial and formal hypotheses at both the "*microscopic*" and "*macroscopic*" levels. By considering the principles of thermodynamics in the design process, architects can create more efficient and sustainable structures that are better suited to their environment.

Figure 2.1. *Climatic Constructal Heterotopia definition. Copyright Lazaros Mavromatidis. For a color version of this figure, see www.iste.co.uk/curulli/heritage.zip*

Within the context of a "*climatic heterotopia*", it is proposed that architecture creates "*large*" thermodynamic systems across multiple scales simultaneously, including the urban, building, envelope and material scales. In this chapter, a brief essay is presented outlining how macroscopic thermodynamic analysis can be applied to any architectural problem. By doing so, accurate spatial hypotheses can be obtained in relation to the surrounding climate, and the evolution of spatial conception can be understood over time.

From a pedagogical point of view, the concept of "*climatic heterotopia*" emphasizes the need to integrate climatic and environmental factors into the architectural design process. By considering the thermodynamic properties of a building and its surrounding environment, architects can create structures that are more energy efficient, sustainable and better adapted to their surroundings. To achieve this, a macroscopic thermodynamic analysis must be performed across various scales, ranging from the urban to the material level. This analysis can provide insight into how energy flows through a building and how it interacts with the surrounding environment. Through the use of simulation tools, architects can evaluate the performance of their designs in different climate conditions and identify opportunities for improvement.

The outcome of this pedagogy is a set of accurate spatial hypotheses that can guide the design process. By understanding the thermodynamic behavior of a building, architects can make informed decisions about the building's form, orientation, materials and systems. This approach enables architects to create structures that are not only esthetically pleasing but also efficient and sustainable. In summary, the integration of thermodynamic principles into architectural design represents a promising approach to create structures that are better adapted to their surroundings. By considering the interaction between a building and its environment at a macroscopic level, architects can create spaces that are comfortable, efficient and sustainable.

2.2. Indeterminacy, autonomy, thermodynamics: the concept of constructal "*climatic heterotopias*"

Contemporary and past architectural conceptions within the context of climate change are dependent on hypotheses and measurements made at the "*microscopic*" level, which involve spatial materiality and interactions with the surrounding environment. At the macroscopic level of "*large*" systems, hypotheses and estimations-simulations are employed to complete this set, alongside subjective and/or indeterminate hypotheses that help imagine the lifecycle of a building/spatial subject and the spatial transitions from the microscopic to the macroscopic level. Throughout the history of architecture, indeterminacy has been directly linked to the nature of architectural conception across multiple levels, illustrating the dimension of architecture that cannot be quantified, measured or fully understood (Picon 2010).

Thermal indeterminacy, also known as "*thermodynamic indeterminacy*", is a phenomenon that has significant implications for architectural design. It refers to the inability to predict with certainty the macroscopic behavior of a thermodynamic system, even if all of its microscopic properties are known. The thermodynamic state of a system may be described by macroscopic variables such as pressure, temperature and volume, but the exact values of these variables cannot be determined with certainty (Callen 1985; Kondepudi and Prigogine 2014).

In architectural design, this means that there will always be some level of uncertainty associated with the performance of a building, particularly as it relates to its thermal behavior. While simulation tools and other technologies can help mitigate this uncertainty to some extent, they cannot completely eliminate it. As such, architects must take into account the inherent indeterminacy of the thermodynamic properties of their designs and incorporate this into their decision-making processes (Kolarevic 2010).

In summary, the indeterminacy inherent in the thermodynamic properties of architectural designs must be taken into account when considering the performance and behavior of buildings, particularly in the context of climate change. While this uncertainty cannot be fully eliminated, it can be managed through the use of simulation tools and other technologies that help predict the behavior of buildings in different scenarios (O'Brien and McGee 2013).

This arises because of the statistical nature of thermodynamics, where the behavior of a large number of microscopic particles, such as molecules or atoms, is described in terms of statistical averages (Leff and Rex 2003). These averages are subject to fluctuations due to the random motion of individual particles, which can lead to uncertainties in the macroscopic properties of the system (Landsberg 1984). Thermodynamic indeterminacy is closely related to the concept of entropy, which is a measure of the degree of disorder or randomness in a system. The higher the entropy of a system, the greater the thermodynamic indeterminacy, since the system is more likely to be in a state that cannot be predicted with certainty (Hill 1987).

However, the process of architectural conception involves the spatialization of multiple thermodynamic states within a single thermodynamic system due to the nature of the discipline. In classical thermodynamics, a thermodynamic state of a system refers to its condition at a specific time and it is fully characterized by a suitable set of parameters, also known as state variables, state parameters or thermodynamic variables (Ben-Naim 2008). A thermodynamic system, on the other hand, is a body of matter and/or radiation that is spatially confined by boundaries, such as walls or envelopes, with defined permeabilities, which separate it from its surroundings (Ben-Naim 2011).

Based on the second law of thermodynamics, there exists an infinite array of potentialities that can be expressed spatially and temporally as micro-states (Denbigh 1971). Micro-states represent the various possible arrangements of spatial position and energy distribution at a particular thermodynamic state (Ben-Naim 2011). The concept of micro-states deals with the spatial distribution and energy position within a system (Denbigh 1971). In the realm of architectural conception, the number of micro-space generations within the overall thermodynamic system plays a crucial role in determining how the passage between the different existing thermodynamic scales is envisaged.

Thus, it follows that architecture can be regarded as a flow system that must exhibit a continuous evolution in accordance with Adrian Bejan's constructal law, which stipulates that for a flow system to endure over time, it must develop in a manner that facilitates easier access to its currents (Bejan 2015; Bejan and Lorente 2013). The constructal law may serve as a blueprint for generating configurations and designs and provides a thermodynamic interpretation of the requisite "*autonomy*" that architecture requires to "*evolve*" and adapt to its current and future environment. As architecture is concerned with the generation of spatial configurations and design shapes, it must possess a degree of autonomy at both the macroscopic and microscopic scales to endure over time and attain the status of heritage (Callen 1985).

Ergo, the "*thermodynamics of architecture*", requires the integration of various hypotheses and measurements across multiple scales. The conception of architectural space involves different approaches to comprehending the overall thermodynamics of architecture, depending on the level at which hypotheses are posed by the designer/architect. The question of coherence arises from ad hoc thermodynamic investigation and the potential degrees of freedom in their establishment. The advancement of thermodynamics and the constructal law (or the renewal of classical thermodynamics) must be situated within this framework. In order to explore potential avenues for detailed research, a brief analytical method is provided in the conclusion section of this chapter.

In the context of architecture and thermodynamics, it is essential to recognize that architecture operates within complex and dynamic systems. The built environment is continuously interacting with its surroundings, and it is shaped by the forces and energies that act upon it. As such, the understanding of thermodynamics in architecture must account for the multi-scale nature of these interactions. Architects and designers must consider the various scales of interaction from the microscopic level of spatial materiality to the macroscopic level of large-scale systems. At the microscopic level, the focus is on the spatial distribution and energy position of the building materials, while at the macroscopic level, the concern is with the overall behavior of the building as a system within its environment.

To achieve coherence in architectural design, it is necessary to balance the various hypotheses and measurements made at these different levels. This involves considering the possible degrees of freedom in the establishment of the design and investigating the ad hoc thermodynamic investigations that contribute to its development. One promising avenue for the progress of thermodynamics in architecture is through the constructal law. This law provides a framework for understanding the evolution of flow systems and the necessary autonomy required for systems to adapt to their environments. By considering the thermodynamics of architecture in this way, designers and architects can better understand how their designs will interact with the environment and adapt over time. To advance this understanding, further detailed research is needed. An analytical method could be developed to help architects and designers balance the various hypotheses and measurements made at different scales and to identify the degrees of freedom in design. Through this research, we can gain a better understanding of how the principles of thermodynamics can be applied to architecture and how we can create more sustainable and adaptive built environments.

In the subsequent sections, this chapter will delineate the pivotal points of the discourse surrounding the thermodynamic constructal aspect of "*climatic heterotopias*" and the dynamic conversion or creation of architectural heritage through thermodynamics, while also examining the role of positionality and standpoint in shaping architecture and heritage. Furthermore, this paper will consider the possibility of analyzing individuals or social groupings as fluids that can be thermodynamically evaluated to produce novel and intricate architectural designs. Additionally, this paper will investigate whether specific formalistic project choices can be assessed using constructal thermodynamics at a macroscopic level to provide superior access to the truth compared to others. Finally, this chapter will explore the extent to which a synergistic methodology that combines architecture and thermodynamics can generate knowledge to create a new heritage, and whether the new knowledge produced through this approach may be incommensurable due to "*epistemic chasms*".

In conclusion, the intricate issues surrounding the intersectionality and mutual construction of theoretical and scientific divisions with regard to heritage and climate change can be resolved as a classic thermodynamic problem that encompasses the interrelationships of position, cultures and values. Drawing on Castoriadis' radical ontology of indeterminacy and related works (Mavromatidis 2019, 2023), I propose an indeterminate continuum of thermodynamic frameworks for architectural conception, particularly in the creation and/or conversion of architectural heritage, between the complete determinacy of traditional spatial conception and the absolute indeterminacy of the traditional conception of nothingness (from the perspective of an egoless state of being in which we fully recognize our own small part in the cosmos), which is subject to reprimand by contemporary pedagogic and scientific systems.

2.3. Macroscopic and microscopic thermodynamics of architectural conception: entropy and evolution

Currently, mainstream research in applied thermal engineering for buildings predominantly relies on microscopic-level hypotheses and measurements related to the envelope and materials, utilizing classical concepts of mechanics such as position, mass, speed, forces, kinetic and potential energy, and Newton's laws (Lambert 2002). In contrast, my proposal for a macroscopic thermodynamic analysis of an architectural problem at an early design stage involves defining quantities such as temperature, quantity of heat and entropy, and their respective properties, which are predetermined (Lambert 2002). Statistical hypotheses may supplement this set of parameters and introduce a framework for transitioning from the microscopic to macroscopic level, taking into account the extensivity property, ergodic postulate, discernible character of individual building components and the definition and equiprobability of microscopic states (Lambert 2002). Contemporary thermodynamics considers the number of microscopic states in the overall system to be crucial in determining the transition between levels (Lambert 2002). In order to externalize architecture through a thermodynamic lens, it is essential to incorporate these various hypotheses and measurements cohesively.

Understanding the crucial dual thermodynamic definition of entropy ($dS = dQ/T$ – Clausius's equation, as well as the statistical definition, $S = kBln\Omega$ – Boltzmann's equation (Boltzmann 1896, 1964)), is a persistent challenge in order to generate architectural space from thermodynamic space (Lechner 1999). Failure to address the challenge of connecting the concept of entropy to heat in a meaningful way results in trivial theories and concepts that lack specificity and objectivity. To overcome this challenge, a dual scientific definition of entropy must be utilized in a methodology that integrates it as a creative tool in architectural design. The second principle of thermodynamics necessitates that any architectural concept must address the inevitable evolution of time and shape from its inception in the mind of the creator (Lambert 2002). Architecture, as a spatial science, shapes time and therefore any architectural entity, be it an object, building, or city, can be viewed as a complex thermodynamic system that is constantly transforming and evolving under changing thermodynamic conditions. This perspective emphasizes the importance of considering the interplay between thermodynamics and architecture in order to fully understand the nature of architectural creations.

Given that architecture is concerned with the evolution of space and time, it is reasonable to assume that fundamental architectural concepts and quantities are initially formed at a macroscopic level, although they share similarities with those that arise at the microscopic level. Some scholars have questioned how time emerges at the macroscopic level from phenomena at the microscopic level, which is not associated with time at that level (Sciamanda 1969).

However, it is hypothesized that the same key concepts used at the macroscopic level are also employed when investigating energy at microscopic scales. In addition to conventional concepts of the microscopic level, temperature, pressure, quantity of heat and entropy are included, along with their a priori properties, which are synthetically expressed in the principles of thermodynamics (Kozliak 2007, 2009; Sciamanda 1969). Architectural tools are then used to explore how these concepts are linked to those derived from the microscopic level and how they are translated into architectural space with specific dynamic thermal characteristics.

It would be a fallacy to approach architecture solely through an ascending or descending methodology, whether by summation or averaging processes. Although thermodynamics provides useful concepts that are scientifically defined independently at the microscopic level, for instance, those applicable to sophisticated building components, we must acknowledge that during the conceptual phase, we ignore almost everything at this level with respect to the actual estimated values of quantities concerning heat transfer. This is due to our reliance on a virtual space of elaborate and time-consuming simulations. We cannot realistically simulate the countless microscopic particles or the information of a mechanical nature (such as positions, mechanical resistance, velocities) usually associated with macroscopic architectural forms. Instead, the macroscopic level on which we conceive architectural and thermodynamic shapes dictates the average quantities to be computed on the microscopic level (e.g. the thermal resistance of the building envelope is derived from the thermal properties of the constitutive elements). Thus, part of the information flows from the macroscopic to the microscopic level. The universal principle of energy conservation, which links the two levels, permits this transfer while the finer description of microscopic quantities necessitates statistical postulates (Kozliak 2004).

In summary, conventional applied thermal engineering for buildings is a reductionist approach that focuses on the thermal evaluation of individual building components, but this approach fails to provide a holistic understanding of the overall system. This reductionism can be compared to the approach in quantum mechanics, where the quantification of energy levels of elementary particles arises from boundary conditions on the entire system and is not an intrinsic property of each particle. Therefore, a thermodynamic investigation of architectural space must adopt a holistic approach that considers the properties and interactions of all building components. While the properties of individual components can be estimated from macroscopic calculations, assuming their lack of interactions, it is crucial to consider the thermodynamic potential of the overall shape. Furthermore, the holistic approach in architectural design and thermal engineering is not only necessary from a scientific perspective but also from a practical standpoint. Buildings are complex systems that involve the interaction of various components and subsystems, such as walls, roofs, HVAC systems and occupants. Neglecting the interactions and feedback mechanisms between these components can lead to suboptimal designs and even result in energy waste and increased carbon footprint. Thus, a holistic approach that considers the building as a whole and the interactions between its components can result in more energy-efficient and sustainable designs. Moreover, the incorporation of advanced control strategies and machine learning algorithms that take into account the holistic nature of the building can further optimize its energy performance and reduce its environmental impact.

2.4. Conclusion

In summary, the thermodynamic concepts of heat, internal energy, entropy and directional evolution are integral to the design of a building, as they arise from a division of scales and a sharing of knowledge between these scales. At the conceptual design stage, architects must consider the overall thermodynamics of the building in relation to small-scale heat exchanges and define thermodynamic quantities for the entire volumetric shape. By applying the constructal law, architects can identify the direction of volumetric evolution at both the macroscopic and microscopic scales. However, this raises the important question of how to reconcile assumptions made at two independent levels without encountering scientific contradictions (MacDonald 2006). The answer to this question warrants further investigation and may be explained by various modes of explanation.

Furthermore, a thermodynamic approach to architectural design could also help in the optimization of energy efficiency and sustainability. By considering the overall thermodynamic potential of a building shape, designers can identify areas of improvement in terms of heat transfer and energy consumption. This can lead to the creation of buildings that not only meet the functional needs of their occupants, but also have minimal impact on the environment. Moreover, a holistic approach to building design can also lead to the creation of esthetically pleasing structures that are both functional and energy efficient. Therefore, a thermodynamic approach to architectural design can have significant implications not only for the scientific understanding of building design, but also for the wider societal and environmental context in which architecture operates. The following are the key steps involved in conducting a coupled architectural and thermodynamic investigation:

– identify the scales of investigation and the boundaries of the system being studied;

– define the relevant thermodynamic quantities and parameters at the macroscopic and microscopic levels;

– develop a conceptual model of the architectural form, taking into account both the macroscopic and microscopic scales and the thermodynamic properties and interactions;

– conduct simulations and experiments to test the performance of the proposed architectural form under various environmental conditions;

– analyze the results of the simulations and experiments, drawing conclusions about the thermodynamic and architectural characteristics of the form;

– refine the conceptual model and repeat the simulation and testing process as necessary to achieve the desired performance.

It is important to note that this investigation requires an interdisciplinary approach, involving expertise in architecture, thermodynamics and related fields. The process may also involve the development of new tools and methods for modeling and simulation, as well as the integration of multiple sources of data and information. Ultimately, the goal is to create architectural forms that are both esthetically pleasing and energetically efficient, contributing to a sustainable and liveable built environment.

At the heart of the coupling between architecture and thermodynamics lies the interdependence of the hypotheses made at both the microscopic and macroscopic levels. This interdependence implies that the two sets of hypotheses are not in competition, but rather complementary, as they relate to different degrees of freedom of the building system. At the same time, these hypotheses are recursively related, in that they are posed in opposition to one another, forming a part of the general thought process, which is not specific to either architecture or thermodynamics. The evolution of architectural thinking and its products relies on relational thinking, as everything we define refers to everything else that is left out (Styer 2000). Moreover, as *climatic constructal heterotopias* are evolutive and governed by the constructal law of thermodynamics, evolution will repair all the relationships that are not well thought out and investigated during the early design stages.

However, during the evolution of the conceptual product, it is possible to encounter incompatibilities between the hypotheses adopted at the early design stages. For instance, there may be a discrepancy between the entropy function constructed at the macroscopic level based on the quantity of heat and temperature, and the one based on statistical hypotheses regarding the probabilities of different states (Leff 2012). Nevertheless, applying the constructal law during the upstream phase will help ensure that the correct direction of evolution is chosen from a thermodynamic point of view. Thus, it becomes clear that the hypotheses made at both the microscopic and macroscopic levels are not independent, but are rather equivalent in a certain sense, and that their recursive relations provide a framework for understanding the coupling between architecture and thermodynamics (Kittel 2004). Furthermore, a coupled architectural and thermodynamic investigation can be a useful tool for identifying and addressing issues related to sustainability and energy efficiency in the built environment. By taking a holistic approach that considers both the macroscopic and microscopic scales, architects and engineers can better understand the interactions between building components and their impact on the overall energy performance of the building. This can lead to more informed decision-making during the design phase, as well as the identification of opportunities for optimizing energy use and reducing environmental impacts throughout the building's lifecycle. As such, a holistic thermodynamic approach to architectural design can play a key role in promoting sustainability and supporting the transition toward a more sustainable built environment.

To further enhance this proposal, it is important to note that the application of constructal thermodynamics in architectural design allows for a more systematic and holistic approach. By considering the thermodynamic properties of the building as a complex system, it is possible to identify the most efficient and effective design solutions that take into account not only the internal heat exchange, but also the external environmental factors. Moreover, by incorporating the constructal law, which states that "*all flow systems evolve in a way that maximizes flow access*", it is possible to identify the direction of the volumetric evolution of the building shape at both macroscopic and microscopic scales. This not only ensures the thermodynamic viability of the design, but also contributes to the esthetic and functional aspects of the building. Therefore, by combining the principles of thermodynamics and architecture, constructal thermodynamics offers a powerful tool for sustainable and innovative design solutions that prioritize both the human and environmental factors.

On arrival, as it is common in architectural theory, there is the possibility of theoretical pluralism. Regarding architectural design, no more than in any other science, the magnitudes and laws of thermodynamics are strictly "*imposed*" by reality and experience. There is arbitrariness during the conceptual scale since the project is conceived in an intelligible word until the passage into the construction phase. However, we then accommodate in different ways by applying constructal thermodynamics: by internal degrees of freedom, designations (with uncertainty) and the many modalities of what we call "*interpretation*" (Styer 2000). A climatic constructal heterotopia is an evolutive dynamic concept that offers many degrees of freedom during architectural conception.

2.5. References

Baker, D. (1997). General chemistry, 5th edition. *J. Chem. Educ.*, 74, 1049.

Bejan, A. (2015). Constructal law: Optimization as design evolution. *J. Heat Transf.*, 137, 061003.

Bejan, A. and Lorente, S. (2013). The constructal law of design and evolution: Physics, biology, technology, and society. *J. Appl. Phys.*, 113, 151301.

Ben-Naim, A. (2008). *Entropy Demystified: The Second Law Reduced to Plain Common Sense*. World Scientific, Singapore.

Ben-Naim, A. (2011). Entropy: Order or information. *J. Chem. Educ.*, 88, 594–596.

Boltzmann, L. (1896). *Vorlesungen über Gastheorie*. JA Barth, Leipzig.

Boltzmann, L. (1964). *Lectures on Gas Theory*, translated by Brush, S.G. University of California Press, Berkeley.

Callen, H. (1985). *Thermodynamics, and an Introduction to Thermostatics*, 2nd edition. Wiley, Hoboken.

Denbigh, K. (1971). *Principles of Chemical Equilibrium: With Applications in Chemistry and Chemical Engineering*, 3rd edition. Cambridge University Press, New York.

Hill, T.L. (1987). *An Introduction to Statistical Thermodynamics*. Dover Publications, Mineola.

Kittel, C. (2004). *Elementary Statistical Physics*. Courier Corporation, North Chelmsford.

Kolarevic, B. (2010). *Architecture in the Digital Age: Design and Manufacturing*. Spon Press, London.

Kondepudi, D. and Prigogine, I. (2014). *Modern Thermodynamics: From Heat Engines to Dissipative Structures*. John Wiley & Sons, New York.

Kozliak, E.I. (2004). Introduction of entropy via the Boltzmann distribution in undergraduate physical chemistry: A molecular approach. *J. Chem. Educ.*, 81, 1595.

Kozliak, E.I. (2007). Consistent application of the Boltzmann distribution to residual entropy in crystals. *J. Chem. Educ.*, 84, 493.

Kozliak, E.I. (2009). Overcoming misconceptions about configurational entropy in condensed phases. *J. Chem. Educ.*, 86, 1063.

Lambert, F.L. (2002a). Disorder – A cracked crutch for supporting entropy discussions. *J. Chem. Educ.*, 79, 187.

Lambert, F.L. (2002b). Entropy is simple, qualitatively. *J. Chem. Educ.*, 79, 1241.

Landsberg, P.T. (1984). *Thermodynamics and Statistical Mechanics: An Intermediate Level Course*. Oxford University Press.

Lechner, J.H. (1999). Visualizing entropy. *J. Chem. Educ.*, 76, 1382.

Leff, H.S. (2012). Removing the mystery of entropy and thermodynamics – Part V. *Phys. Teach.*, 50, 274–276.

Leff, H.S. and Rex, A.F. (2003). *Maxwell's Demon 2: Entropy, Classical and Quantum Information, Computing*. CRC Press, Boca Raton.

MacDonald, D.K.C. (2006). *Introductory Statistical Mechanics for Physicists*. Courier Corporation, North Chelmsford.

Mavromatidis, L. (2019). Constructal macroscale thermodynamic model of spherical urban greenhouse form with double thermal envelope within heat currents. *Sustainability*, 11, 3897.

Mavromatidis, L. (2023). Constructal evaluation of polynomial meta-models for dynamic thermal absorptivity forecasting for mixed-mode nZEB heritage building applications. *Energies*, 16, 429.

O'Brien, W. and McGee, W. (2013). *Predicting Future Climate Changes: Scientific Challenges and Technological Advances*. Springer, New York.

Picon, A. (2010). *Digital Culture in Architecture*. Birkhauser, Basel.

Sciamanda, R.J. (1969). Expansion of available phase space and approach to equilibrium. *Am. J. Phys.*, 37, 808–809.

Styer, D.F. (2000). Insight into entropy. *Am. J. Phys.*, 68, 1090–1096.

3

Training for Cultural Heritage Protection: The STRENCH Summer Schools

Natural and man-made hazards, anthropogenic effects and extreme events related to climate change increasingly threaten cultural and natural heritage both in Europe and worldwide (Bonazza et al. 2018). Within the framework of the Interreg CENTRAL EUROPE project STRENCH, which itself is based on the predecessor project ProteCHt2save, innovative preservation and safeguarding measures and approaches were developed. They include ready-to-use solutions for assessing climate change effects and for defining strategies for the protection of cultural and natural heritage at risk. An explicit aim of the project was to empower local stakeholders to prepare for and act during an emergency and to facilitate the recovery of cultural and/or natural heritage. For this, two summer schools aimed at local, regional and national authorities, civil protection agencies and future generations were an integral part of the project from the very beginning. The project itself was proposed before the Covid-19 pandemic hit in 2020, thus the summer schools were planned in the 1-week on-site format, which the following two organizing institutions had perfected over the years – the Institute of Atmospheric Sciences and Climate of the National Research Council of Italy and the Centre for Cultural Property Protection of the University for Continuing Education Krems, Austria. However, the pandemic meant that a change in teaching approach became necessary – the two summer schools conducted their classes online. It was always clear to the organizers that the practical hands-on approach that lay at the core of former on-site trainings had to be integrated into the online versions as well. The aim of this chapter is to (1) outline the project STRENCH in the details necessary to put the summer schools into context, (2) describe the methods and approaches of the two STRENCH online summer schools from 2020 and 2021 in detail, and (3) to

---

Chapter written by Anna KAISER.

highlight on-site follow-ups aimed at fostering cooperation and coordination between different entities paramount for cultural and natural heritage protection.

3.1. The STRENCH project

STRENCH was a so-called capitalization project, aimed at capitalizing predecessor projects in which partners of the consortium had participated in and further developed and finetuned products and outputs of earlier projects. The project's main objective was to raise the effectivity and efficiency in cultural heritage protection by capitalizing nine previous EU projects and bringing selected results of these projects to the next stage. The projects capitalized were the Interreg Central Europe projects ProteCHt2save, RUINS, HICAPS, BhENEFIT, the H2020 projects HERACLES and SHELTER, the FP6 project Noah's Ark and the FP7 project Climate for Culture. STRENCH focused on the development, testing, and finetuning of ready-to-use solutions for local authorities, such as the WebGIS tool, which was started up in ProteCHt2save, methods for ranking the vulnerability of cultural and natural heritage assets, and strategies for risk minimization and for cultural heritage protection (Bonazza et al. 2021; Cacciotti et al. 2021).

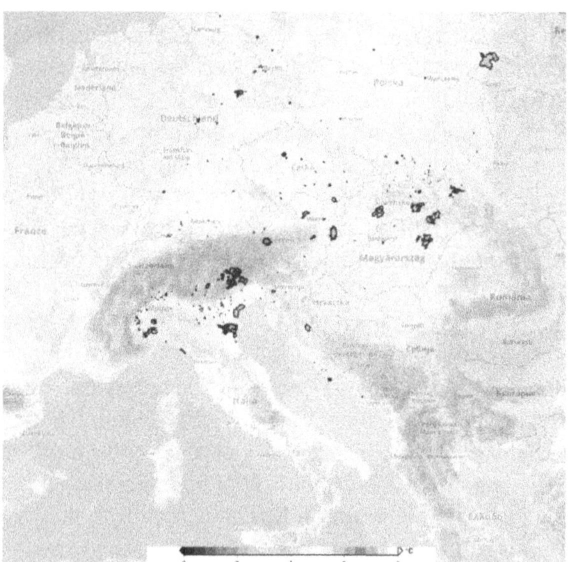

Figure 3.1. *Screenshot from the ProteCHt2save and STRENCH WebGIS tool. The dark blue shapes are the UNESCO World Heritage site in Central Europe (https://www.protecht2save-wgt.eu/). For a color version of this figure, see www.iste.co.uk/curulli/heritage.zip*

Increasing the know-how of local stakeholders on how to effectively prioritize in cultural heritage protection was another crucial aspect of the project. To that end, each of the seven partner countries selected one pilot site from the categories cultural landscapes, historic parks, archaeological sites and small ruined villages in mountains in coastal areas. These pilot sites were pivotal for testing the STRENCH products and for the two training actions in risk management and safeguarding cultural heritage and landscapes conducted as part of the project. The seven strategies for sustainable risk management for cultural heritage developed for the pilot sites contribute to Priority 4 of the Sendai Framework for Disaster Risk Reduction, which is enhancing disaster preparedness and building back better. Additionally, STRENCH supported this priority by developing hazard maps, applying satellite services (for the WebGIS tool, which is based on Copernicus data), creating a measure-oriented database on the criticalities of cultural heritage and landscapes and by establishing intervention priorities to strengthen the resilience of cultural and natural heritage. The main threats considered were heavy rain, floods (in large basins), flash floods, landslides, fire (due to drought) and storms.

To enable caretakers of cultural and natural heritage as well as local and regional administrative bodies and policy makers, two international summer schools were an integral part of the project's set-up. The summer schools, which took place in 2020 and 2021, thus in years 1 and 2 of the project's lifetime, were also a great opportunity for the project staff to test out their ideas and products with future end users. They were therefore able to tailor the outcomes of STRENCH to their needs.

3.2. The STRENCH online summer schools (2020 and 2021)

The traditional mixture of the organizers' summer events consists of theoretical lectures and working groups in which participants work on selected case studies, table-top or live exercises. This approach had been very successfully used in the organization of both the "Cultural Property Protection Summer School", organized by the Centre for Cultural Property Protection of the University for Continuing Education in Krems and the "ENVIMAT Summer School", organized by the Institute of Atmospheric Sciences and Climate of the National Research Council of Italy. Due to the pandemic, the challenge for both the 2020 and 2021 STRENCH summer schools was to keep the highly practical approach while conducting the summer schools in a purely online format.

The setup of the two summer schools was slightly different. In 2020, a pre-learning week featured four video lectures as an introduction to heritage protection from different angles, which included an introductory lecture on cultural heritage protection, dealing with the following subjects: why our cultural heritage needs protection; the military cultural property protection perspective in the 21st century; the Italian model for cultural heritage protection in the case of natural disasters – the so-called "Blue Helmets for Culture", and cultural heritage protection design with the identification of threats, risk and vulnerability assessments. This was followed by live online lectures in the second week, which, to reduce online fatigue, were reduced to one or two lectures per afternoon. The topics treated were more in depth on cultural heritage protection and climate change, thus the intrinsic STRENCH topics: climate extremes in a changing climate; cultural heritage at risk due to the impact of extreme climate, with a focus on the role of satellite observations; the Copernicus satellite program in support of the safeguarding of cultural and natural heritage at risk; vulnerability of cultural heritage as challenges and opportunities for resilience building; risk mapping for the protection of cultural heritage exposed to climate extremes; characterization, damage and protection of building materials for cultural heritage; managing cultural heritage at risk in a changing environment. These theoretical inputs were followed by a practical session on the WebGIS tool, held by the developers from the Institute of Atmospheric Sciences and Climate of the National Research Council of Italy. The final day of the second week was dedicated to the practical testing of this tool by the participants themselves and giving feedback to the developers regarding user friendliness and usability for their specific regions in (Central) Europe. The WebGIS tool was the central product tested in the 2020 STRENCH summer school. At this phase of the project's lifetime, the other results of the project were still in development and could not be tested extensively until the 2021 summer school.

The setup of the 2021 summer school was slightly different: the 1-week program featured full timetables every day, starting with a theoretical input session in the morning, followed immediately by work on case studies in the afternoon, using the tools and methods presented in the morning sections by the developing project partners themselves. For the 2021 summer school, the participants had to choose and prepare a case study to work with during the whole week. They were encouraged to pick a cultural or natural landscape following the STRENCH pilot sites criteria that was well known to them. Each day after the theoretical lectures they used the presented tools on their case study, thus developing protective measures for their sites and testing the STRENCH results (which at that time were still a work in progress) and giving immediate feedback to the developers. The focus of the 2021 summer school lay more on cultural landscapes than cultural heritage. The lectures included climate change challenges for cultural landscapes inter alia impacts and approaches for protection; the Copernicus program again played a huge role, being assessed for the safeguarding of cultural and natural heritage at risk, as did the WebGIS tool in its then current form. This first lecture block was followed by an afternoon of working on individual case studies regarding climate change challenges. To this end, participants with similar case studies were grouped into syndicate working groups to allow them to collaborate and exchange ideas and visions in Zoom breakout rooms. The second lecture block included the impact of natural and man-made threats to cultural heritage, including preparations, vulnerability assessments for cultural heritage protection and, as a special topic, the challenges and perspectives of valorization and the conservation of underwater heritage. This block was followed by syndicate work on the vulnerability of heritage and preventive conservation. The third block was dedicated to assessing threats and managing risks for cultural heritage, including self-caused climate change in cultural landscapes and how to assess their vulnerability with rapid scoring tools; a special lecture was contributed from the northern United States, presenting the "Host Nation Fort Drum" and how cultural landscapes are protected in northern New York in Fort Drum. The final syndicate working afternoon dealt with risk assessment and management. On the following day, the participants of the summer school received further up-to-date research inputs into the topic through the 1st International STRENCH Conference on "strengthening resilience of cultural heritage at risk in a changing environment", which was also conducted online. The 2021 summer school ended with an open presentation of the case studies, given by the participants who developed them during the week.

More than 100 participants from literally all over the world joined the two STRENCH summer schools. The high number and the geographical origins of the participants were a benefit of the online setting, which was in place due to the pandemic (Europa Nostra 2020).

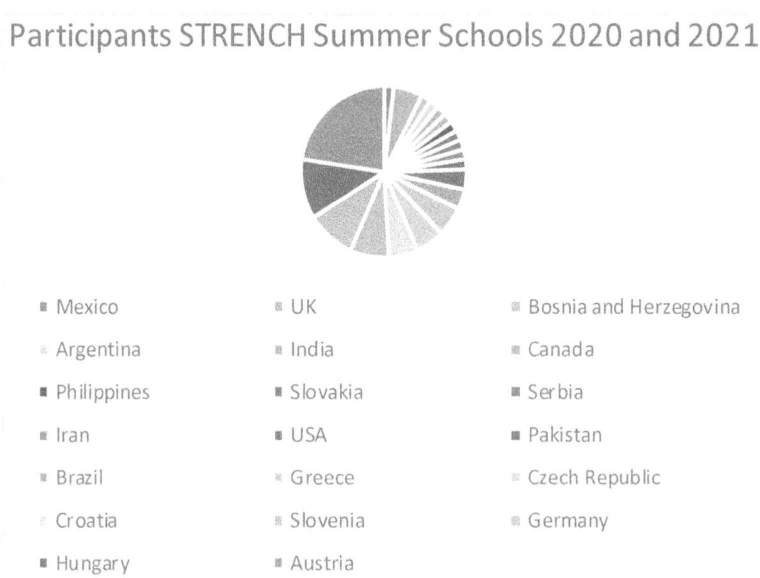

Figure 3.2. *Participants of the 2020 and 2021 STRENCH summer schools. For a color version of this figure, see www.iste.co.uk/curulli/heritage.zip*

They tested the project results tailored to boost the resilience of Europe's cultural and natural heritage to climate change, and used them on specific case studies, thus already preparing sites for future challenges, and adding vital expertise to their portfolio of cultural and natural heritage protection. The project staff of STRENCH got valuable feedback from the participants on the WebGIS tool, the hazard maps developed in STRENCH, the methodology for vulnerability ranking and the strategies for disaster risk reduction, which in addition to the seven pilot sites in the project were also tested for numerous additional sites from all over the world during the 2021 summer school. The set-up of the two summer schools was evaluated quite favorably by the participants, with the 2021 summer school being more focused on practical case study work by the participants themselves. The next step, and this was highlighted by the participants in the evaluations, would be a practical session on site, as was offered several times already by the Centre for Cultural Property Protection of the University for Continuing Education in Krems in the past, taking the form of table-top or live exercises, aiming to facilitate the collaboration of cultural heritage experts with emergency responders in calamitous events threatening cultural and natural heritage. To complete this chapter, on-site exercises as a possible follow-up on online summer schools are described in the last section.

3.3. On-site follow up exercises – the example of "TRITOLIA18"

The exercise TRITOLIA18 was a combination of table-top and a live exercise in which the scenario involved an earthquake that impacted the eastern part of the country, destroying a castle listed as a UNESCO World Heritage site. The exercise was called TRITOLIA18 as it took place in the Austrian Armed Forces' disaster relief training area known as "Tritolwerk", a former ammunition factory dating back to before World War I. The principal purpose of this training exercise was to facilitate the cooperation of the primary actors involved in heritage protection in a crisis – the "heritage side" and the "emergency response side" – and to create interfaces for successful collaboration. The focus of instruction was on heritage professionals, exposing them to how emergency responders (in this case, the Austrian Armed Forces) work. This included acquainting them with the strict decision-making processes and related procedures typical of military entities and educating heritage professionals about how to best communicate critical information regarding cultural heritage to the right person at the right time. It should be noted that this exercise did not aim to train heritage professionals or emergency responders on how to treat impacted cultural heritage and materials. However, the treatment of these mock cultural heritage items was by no means haphazard or blatantly wrong.

On the first day of the exercise, the participants planned the deployment of cultural heritage experts while embedded in urban search and rescue platoons. In other words, during this first stage, the heritage professionals were tasked with working in tandem with battalion staff officers, but at this stage deployed as staff officers and not as heritage experts. On days two and three of the exercise, the heritage sector participants took over the role of subject matter experts, following the plans they helped to develop during day one. They conducted reconnaissance in the affected area and recovered cultural heritage on-site in cooperation with an urban search and rescue platoon, again, as planned during day one of the exercise. This mixture of table-top training for the planning and decision-making components followed immediately by the hands-on recovery of cultural heritage and application of the participants' inherent expertise in correctly handling cultural heritage items was highly appreciated.

Figure 3.3. *On-site live exercise TRITOLIA18 in Austria (DBU/Schramm 2018). For a color version of this figure, see www.iste.co.uk/curulli/heritage.zip*

Expanded training in military staff work is typically unavailable to civilian heritage professionals who do not normally need this type of expertise in executing their day-to-day professional duties (Schramm 2019, p. 13). Time pressure was thus immanent in the exercise – both in the scenario itself on the default setting and outside the scenario, as the time allocated for pre-learning was also limited. Staff work and decision-making processes were aligned to the format used by the

Austrian Armed Forces since they supported the exercise, but also as a means of benefitting their personnel as they were able to train their teams in cooperating with cultural heritage protection. It is also important to note that with only slight modifications the same system is used in civilian disaster management on the national level (Bundesministerium für Inneres 2007).

One key challenge in cultural heritage protection is that the varied experts involved do not work together consistently and are, therefore, often unaware of what the other parties need to facilitate the protection of movable or immovable cultural heritage. The two sides involved in the protection of cultural heritage are often the heritage specialists (mostly civilian experts who rarely have military or disaster management experience) and the military or first responders who are involved in managing man-made or natural catastrophes that threaten cultural heritage. These experts, conversely, rarely have a heritage background. To establish effective cooperation in an emergency, both sides must understand how their respective counterparts operate and what they need to be able to cooperate. Military staff are used to well-focused and strictly organized ways of working, which might seem quite alien to someone who has never encountered such procedures. The military and other emergency responders, that is, civil protection and fire brigades, are not likely to change their established and ingrained procedures. Hence, the civilian experts should at least know how the military procedures work and how best to integrate themselves and their knowledge into this rigid system. As a result of this exercise, civilian heritage specialists gained a fundamental understanding of how military personnel work and how to best integrate themselves and their expertise into operational procedures. They were able to communicate and contribute their vital knowledge about the cultural heritage to the right people at the right time and in the right ways.

The TRITOLIA18 exercise was conducted as part of the Interreg CE project ProteCHt2save, for which a novel approach was developed and tested to bridge the above-mentioned gap. This approach takes the form of so-called Cultural Heritage Rescue Teams (CHRT). Ideally, these teams are composed of a mixture of heritage experts and personnel from emergency units or the military, but a team composed solely of heritage experts could do the job as well, provided they understand how the military and emergency units plan and operate (Kaiser 2019).

The STRENCH summer schools have shown that practical approaches to heritage education for climate action can be transferred into a highly productive online setting, much appreciated by participants from different heritage backgrounds. The additional opportunity of providing the participants with an on-site exercise should be given wherever possible. Bringing together participants from heritage and emergency responders' sides has proven to be highly beneficial for both sides when it comes to training for cultural heritage protection.

3.3. References

Bonazza, A., Maxwell, I., Drdácký, M., Vintzileou, E., Hanus, C. (eds) (2018). Safeguarding cultural heritage from natural and man-made disasters: A comparative analysis of risk management in the EU. Publications Office of the European Union, Luxembourg [Online]. Available at: https://op.europa.eu/en/publication-detail/-/publication/8fe9ea60-4cea-11e8-be1d-01aa75ed71a1.

Bonazza, A., Sardella, A., Kaiser, A., Cacciotti, R., De Nuntiis, P., Hanus, C., Maxwell, I., Drdácký, T., Drdácký, M. (2021). Safeguarding cultural heritage from climate change related hydrometeorological hazards in Central Europe. *International Journal of Disaster Risk Reduction*, 63, 102455.

Bundesministerium für Inneres (ed.) (2007). Richtlinie für das Führen im Katastropheneinsatz [Online]. Available at: https://www.bmi.gv.at/204/Download/files/Richtlinie_fuer_das_Fuehren_im_Kat-Eneinsatz_Letztfassung_BF_20200122.pdf.

Cacciotti, R., Kaiser, A., Sardella, A., De Nuntiis, P., Drdácký, M., Hanus, C., Bonazza, A. (2021). Climate change-induced disasters and cultural heritage: Optimizing management strategies in Central Europe. *Climate Risk Management*, 32, 100301.

Europa Nostra (ed.) (2020). Covid-19 & beyond: Challenges and opportunities for cultural heritage [Online]. Available at: https://www.europanostra.org/wp-content/uploads/2020/10/20201014_COVID19_Consultation-Paper_EN.pdf.

Kaiser, A. (2019). Guidelines for Cultural Heritage Rescue Team (CHRT) [Online]. Available at: https://www.interreg-central.eu/Content.Node/D.T3.2.1-Guidelines-for-Cultural-Heritage-Rescue-Team.pdf.

Schramm, H. (2019). Integration von zivilen Akteuren des Kulturgüterschutzes in einen Einsatzstab. Ein Planspiel zur Vermittlung von Kenntnissen zu Funktion und Verfahren. MA thesis, Danube University Krems.

4

Pedagogical Aspects of Conservation-Restoration Practice in the Protection of Architectural Heritage

This chapter considers the pedagogical aspects of the cultural heritage (CH) field. The work of CH specialists often remains hidden from the non-expert observer, but it is of high public importance because it maintains the cultural facade of every society. CH and issues related to activities such as conservation, restoration, reconstruction of CH and museum activity must be included in the educational programs because they are an essential addition to the general culture of modern humans. This chapter focuses on the necessity for a purposeful inclusion of the topic related to CH, both in the earliest educational stages and in higher education. It examines both the upbringing component and the educational component. It also presents various target groups in which pedagogical principles are used. The role of CH in the formation of young people's values has been confirmed in consideration of different target groups depending on the educational goals and the didactic principles applied. In the contemporary world, didactic educational games are becoming more relevant, reaching the popularity of classical joint art workshops and master classes in protecting CH.

4.1. Introduction

The work of restoration, the architect, the art restorer and the museum worker are out of the spotlight. The result of CH professionals is of exceptional public importance, although they remain hidden from the ordinary observer. The product of their work can be seen by all. It becomes a cultural facade of society, making the pedagogical aspects of preserving architectural heritage current for the community. The in-depth knowledge of the specifics of the activities for the protection of CH is an essential addition to the general culture of the education of contemporary

---

Chapter written by Yavor DANKOV, Yakim DEYKOV and Andjela DANKOVA.

humans. Therefore, it is necessary to purposefully involve appropriate didactic practices in the early stages of education and train specialists from other professional fields. Knowledge and attitude to the issues related to the protection of CH contribute to the correct risk assessment and strategic management decisions. Along with joint art workshops and master classes, didactic educational games are becoming more relevant in protecting CH.

The chapter refers to the *pedagogical aspects* of preserving CH. They contain education and upbringing – dialectically linked directions that the science of pedagogy deals with (Comenius and Keatinge 1967). In this chapter, they are presented in a vital link. This chapter is based on axioms of pedagogy and didactics (Comenius and Keatinge 1967; Jůvová and Bakker 2015; Guseva 2018). Examples of didactic principles refer to clarity, consciousness, activity, accessibility, scientificity, systematicity and consistency, permanent assimilation of knowledge, skills and habits, individual approaches, initiative and self-direction, flexibility, adaptability, etc. (Jůvová and Bakker 2015; Hernández 2021). In this regard, among the commonly used approaches for teaching and learning are holistic approaches, responsiveness to learners, learning through play, intentional education, learning environments, cultural competence, continuity of learning and transitions, assessment for learning, etc. (Kreig 2011; Ildiz and Ahmetoğlu 2017). The subject of pedagogy, such as applied science, is the technology of education and upbringing (Ildiz and Ahmetoğlu 2017). The presence of an *educational* and *upbringing component* is a fundamental postulate in pedagogical theory (Moskaliuk 2018). The educational component rests on current scientific facts and proven laws and patterns. Upbringing is an investment in young humans' morals. The aggregate result of good pedagogical practice should be building personalities of a balanced character, educated according to the current level of scientific thought, and brought up according to undoubtably human and humanitarian values.

The concept of CH also refers to the material (architecture, urban planning, sacred objects, artistic artifacts) and intangible heritage (language, customs, traditions, legends, music, folklore). They are subject to study, cataloging, reconstruction and conservation (Germanà and Nicolini 2022; Leila and El-Bastawisy 2022). The aggregate activity, named preservation of CH, contains many aspects: scientific, pedagogical, artistic, tourist, economic, etc. A good knowledge of history and its material basis (archaeological and cultural artifacts) testifies to the formation of significant personalities with a waking conscience and people with self-confidence and inclusiveness, and a sense of belonging to communities.

The focus of this chapter is on the necessity for purposeful pedagogical work with the whole society from its earliest educational stages, corresponding to the CH problems. Any country striving to be responsible for the CH of its territory, which is part of the world CH, should apply such an educational policy to have actual results in this direction. This chapter presents various *target groups* in which *pedagogical principles* are used. When working with different target groups, depending on the educational goals and *applying the didactic principles* presented, the role of CH in forming values of both the young person and the more mature specialist is affirmed. The Summer Drawing Practicum intended for students from the master's program "Architecture" at the University of Architecture, Civil Engineering and Geodesy UACEG, Sofia, Bulgaria, can be given as an example of applying the presented didactic principles in the second target group. These are the specialists that will become future architects. In the practicum, a specialized educational program is applied in the creation and development, which the authors have contributed to. In modern pedagogy, game methods are becoming increasingly important, which is also valid for the pedagogical aspects of the preservation of CH (Ekin et al. 2023).

4.2. Application of didactic principles in different target groups

The application of didactic principles in the field of CH is presented in the following target groups:

– children and young people – students from the elementary, primary and secondary educational stages;

– students whose disciplinary expertise directly corresponds to the CH issues;

– students whose disciplinary expertise does not directly correspond to the CH issues;

– citizens with extracurricular status.

4.2.1. *Target group 1 – children and young people*

The first target group refers to the young people in the early educational stages. Practical pedagogy includes two directions – *curriculum* and *extracurricular activity* (Khakimova et al. 2022). The curriculum contains appropriate objects – the science of birth, history, geography, arts, etc., within which it is reasonable to consider representative CH (movable and immovable) and achievements of the intangible CH (Barghi et al. 2017). Pedagogical work contains direct messages and indirect suggestions. The early creation of attitude toward CH is associated with the appropriate illustration of the textbooks, the skillful selection of photo material, the selection of literary excerpts in the readings, the choice of songs in music classes and the production of drawings in fine arts classes. Extracurricular activities include excursions, visits to museums, galleries and temples, collective and individual reading, art workshops, classical and electronic didactic games and so on. The upbringing component is mandatory in curriculum and extracurricular activity. The way the information is presented is essential for the learner to perceive this information as attractive and captivating, provoking respect for the old masters and admiration for the greatness of the past. The young are susceptible to positive suggestions, curious and emotional, and love tales, legends and stories (Vangsnes et al. 2012). Art is the most meaningful game in childhood and youth. It is a bearer of sensory knowledge, joy, inspiration, good mood and indirect education.

4.2.2. *Target group 2 – students whose disciplinary expertise directly corresponds to the CH issues*

The second target group refers to students with expertise directly corresponding to the CH problems. Disciplines such as history, archaeology, culturology and tourism cannot be indifferent to the topic. Practical pedagogy includes three directions – curriculum, extracurricular employment (practices, creative workshops) and parallel professional activity.

Students in the disciplinary expertise of conservation and restoration are deeply related to the current CH problems (Figure 4.1). In their qualification and the fundamental ability to paint, there should be sufficient hours related to the art sciences – chemistry, art history, technology, museum work, physics and microbiology. In addition to the qualification to be acquired, they must be brought up in respect and praise for the achievements of the old masters, in artistic modesty, and not to point out their personalities in professional practice (Achille and Fiorillo 2022). An excellent future restorer must have a sense of responsibility and an unmistakable sense of minimal (necessary and sufficient) intervention in restoring original works of art – movable and immovable.

Architecture students are educated to implement new construction and become experts in preserving and reconstructing immovable architectural and archaeological sites. The skillful socialization of the cultural-historical complexes and their adaptation to a modern environment depend on specialists such as architects and urban planners (Embaby 2014; Abouelmagd 2023). When preparing for professional life, the young architect should have a clear idea of the challenges of new construction and the problems of conservation and restoration of immovable cultural monuments before receiving legal qualification to work (Boarin and Martinez-Molina 2022). Teamwork with artist-restorers, archaeologists and specialists in museum work should be improved during education practices and within the framework of targeted creative workshops (Danaci 2015). The Summer Drawing Practicum (in the creation and development of which the authors have contributed) intended for students from the master's program "Architecture" in UACEG, Sofia, Bulgaria, can be given as an example of applying the presented didactic principles in the second target group. During the practicum, students attend various CH sites that they can explore through the creation of a drawing and perform practical tasks to explore the history, problems and importance of CH objects. As a result of completing the practicum, students acquire individual attitudes and respect for CH.

Figure 4.1. *Process of restoration of the frescoes in the temple of St. Nicholas the Miracle Maker Church, Varna, Bulgaria – Painter-restorer Assoc. Prof. Yakim Deykov, PhD. For a color version of this figure, see www.iste.co.uk/curulli/heritage.zip*

Archeology students need real work on the ground as part of their studies. The work of the archaeologist depends on the formation of a correct vision of the historical past, based not on assumptions, but on facts (Attenni and Griffo 2019). The archaeologist discovers ancient artifacts and interprets them in the context of historical theory. The archaeologist should be brought up with impeccable professional ethics, and their historical interpretations should be accurate.

4.2.3. *Target group 3 – students whose disciplinary expertise does not directly correspond to the CH issues*

The third target group refers to students with other disciplinary expertise. In this group, pedagogy specialists are considered the most important as they educate children from a young age. Their activity is not directly related to CH, but they must create an attitude toward CH in children. Suppose the teacher is uninterested, unprepared and indifferent to the CH issues. In that case, their future students (initially educated in the first target group) are seriously affected. Therefore, the public task of pedagogy (to form educated, upbrought and responsible personalities) becomes difficult and almost impossible. Professional ethics is developed based on proper upbringing at an early age. Students in the disciplinary expertise of pedagogy should take a course of training in the professional fields of different types of arts – music, fine arts, theater and literature. Therefore, they will be helpful in their role as educators and methodists.

4.2.4. *Target group 4 – citizens with extracurricular status*

The fourth target group refers to citizens with extracurricular status. They receive information on the CH topic from books, popular programs, guides during thematic excursions and public lectures led by famous specialists or experts. With this target group and the other three target groups, the didactic games or serious games play an essential role (Dankov and Bontchev 2020; Steinrücke et al. 2023; Ekin et al. 2023). Good civic awareness of transient values and the problems related to their conservation is not only a scientific responsibility (Bonfantini and Casonato 2022), but also an important, often underestimated obligation of political entities from all levels – local, national and international policy. In modern pedagogy, lifelong education is promoted (Marchigiani 2022), as well as lifelong upbringing. Literature and other arts, television shows, guide talks and state advertising have a substantial educational, or vice versa, corrupting influence, even in advanced age.

4.3. Architecture and CH

Architecture deserves special attention in the context of CH. Architectural works (not only classic works, but also many recent ones) are cultural monuments. But cultural monuments are often obstacles to ambitious investors, whose attitude makes them “enemies of progress”. Tourism, for its part, is also a large business and should support the “old”. But modern standards and comfort requirements often contradict the “spirit” of old architecture in the tourist product. The environmental theme is also intertwined with the problem. Architects, painter-restorers, historians, lawyers and politicians are resolving these complex cases. Decisions taken at professional and political levels testify to a nation's and society's maturity. Architecture is not just a function carrier, an individual creative expression, or a simple design product – architecture carries wordless messages. The architecture itself possesses an upbringing function. The significant urban development, temple and residential architectural complexes are a monument to the human spirit, human impulses and aspirations of dreams of a harmonious and reasonable world. Literary utopias also have their architectural visions. The great visionaries are also the Renaissance masters who created and painted ideal cities (conceptual urbanism) and carried out remarkable monumental structures – great architectural works that changed the medieval world and are the basis of present-day modernity and postmodernity. It is necessary to design and build architecture with respect for what was created by the ancestors, with praise for history, with the pursuit of stylistic attachment, and with restraint for the temptations of eclecticism and fashion. The architectural product is distinguished by its durability in time, and transitional styles do not apply to it. New construction systems and new materials and tools have their marks on both new construction and application areas, such as restoration and reconstruction. Particular attention should be paid to the specialists' creative tact and professional correctness to achieve architectural authenticity between classic and modern, not only in shape and color, but also in the materials used.

Architecture is a visible image of the globalizing world. The problems of architecture's attachment to ecology are relevant, with the responsibility for protecting the significance of history. This should begin in the student years (or earlier), but not be interrupted by career growth. There is an opportunity for a good balance between new and old, good examples of combining classics and modernism when creating a new architecture in the context of CH.

4.4. The role of didactic games in forming knowledge and attitude toward the CH issues

The possible reduction of human activities (including the most strenuous activities) to playing gives Huizinga (1982) the means to define and justify the existence of Homo Ludens in detail. The person playing is a kind of manifestation of an intelligent person.

Undoubtedly, the game marks human evolution. The game accompanies us not only in childhood, but also in the course of life. Culture and art rely on the stage play. In pedagogy, the game is and should be a powerful methodological instrument. Therefore, the game possesses and inspires competition, emotion and adrenaline. The game creates conditions for indirect education – mastering knowledge and delivering educational messages. Didactic games are a counterpoint to commercial games, and their integrated messages should be responsible and purposeful.

Figure 4.2. *A 3D model of the conceptual object cathedral, built through a modular exposition structure, Arch. Andjela Dankova, PhD*

An example in which didactic principles can be implemented is the author's modular design system for exhibition space, presented in the dissertation of Arch. Andjela Dankova, PhD., "Design concept on creating modular constructions for non-traditional expositional environment and their transformation into a conceptual item" (Andreeva 2020). The developed modular design system for exhibition space separation allows the modules from the system to not be stored passively, but to serve as elements for creating conceptual compositions – objects and environmental art. This form of play is, at the same time, a creative workshop that can unite participants of different ages and qualifications. The act of creating can be a topic of a specialized workshop intended for both children and diverse groups of professionals. The conceptual art, located in the architectural interior or the urban space, makes it possible to provoke participants and viewers to reflect on ecology and CH conservation. The modules can play with the participant's imagination and creativity when creating new conceptual compositions based on a defined topic (Figure 4.2).

Educational video games are prevalent nowadays (Ekin et al. 2023). Interactive behavior and competitive character attract new generations (Ekin et al. 2023). Computer programs and their interfaces are close to the life and conduct of modern humans (Dankov 2019). At the same time, didactic educational video games compete with commercial ones (Dankov 2023). This competition is a considerable challenge for their authors. They are expected to interweave educational and upbringing components, meaningful and fruitful messages, and good morals skillfully. At the same time, the creators of a commercial product rely on primary suggestions and do not always observe ethical norms. Didactic electronic games are an excellent field for creating educational products concerning history, culture, creativity, ancient artifacts, customs, traditions, ethnographic and musical heritage – the whole spectrum of what is defined as CH.

4.5. Conclusion

The in-depth knowledge of the specifics of the activities for the protection of CH is an essential complement to the general culture of the education of modern society. The chapter presented the necessity for purposeful pedagogical work with the community from its earliest educational stages, corresponding to the CH problems. The various target groups in which pedagogical principles were used were successfully presented in the chapter. When working with different target groups, depending on the educational goals and applying the didactic principles presented, the role of CH in forming values of both the young person and the more mature specialist was established. In modern pedagogy, game methods are becoming increasingly important, which is also valid for the pedagogical aspects of the preservation of CH. The knowledge and attitude toward the issues related to the protection of CH contribute to the correct risk assessment and strategic management solutions.

4.6. References

Abouelmagd, D. (2023). Sustainable urbanism and cultural tourism, the case of the Sphinx Avenue, Luxor. *Alexandria Engineering Journal*, 71, 239–261.

Achille, C., and Fiorillo, F. (2022). Teaching and learning of cultural heritage: Engaging education, professional training, and experimental activities. *Heritage*, 5(3), 2565–2593.

Andreeva, A. (2020). Design concept on creating modular constructions for non-traditional expositional environment and their transformation into a conceptual item. PhD Dissertation, Technical University of Sofia.

Attenni, M. and Griffo, M. (2019). Built and destroyed memory: Where does methodology lead? In *Conservation, Restoration, and Analysis of Architectural and Archaeological Heritage*, Inglese, C. and Ippolito, A. (eds). IGI Global, Pennsylvania.

Barghi, R., Zakaria, Z., Hamzah, A., Hashim, N.L. (2017). Heritage education in the primary school standard curriculum of Malaysia. *Teaching and Teacher Education*, 61, 124–131.

Boarin, P. and Martinez-Molina, A. (2022). Integration of environmental sustainability considerations within architectural programmes in higher education: A review of teaching and implementation approaches. *Journal of Cleaner Production*, 342, 130989.

Bonfantini, B. and Casonato, C. (2022). Introduction. In *Cultural Heritage Education in the Everyday Landscape. Digital Innovations in Architecture, Engineering and Construction*, Casonato, C. and Bonfantini, B. (eds). Springer, Cham.

Comenius, J. and Keatinge, M. (1967). *The Great Didactic of John Amos Comenius*. Russell & Russell, New York.

Danaci, H.M. (2015). Aesthetics in cultural landscape and architectural education. *Procedia – Social and Behavioral Sciences*, 191, 190–195.

Dankov, Y. (2019). Conceptual model of user interface design for general architectural framework for business visual analytics. *Proceedings of the 20th International Conference on Computer Systems and Technologies (CompSysTech '19)*. ACM, New York.

Dankov, Y. (2023). DIZU-EVG – An instrument for visualization of data from educational video games. In *Software Engineering Perspectives in Systems*, Silhavy, R. (ed). Springer, Cham.

Dankov, Y. and Bontchev, B. (2020). Towards a taxonomy of instruments for facilitated design and evaluation of video games for education. *Proceedings of the 21st International Conference on Computer Systems and Technologies (CompSys-Tech'20)*. ACM, New York.

Ekin, C., Polat, E., Hopcan, S. (2023). Drawing the big picture of games in education: A topic modeling-based review of past 55 years. *Journal Computers & Education*, 194, 104700.

Embaby, M.E. (2014). Heritage conservation and architectural education: "An educational methodology for design studios." *HBRC Journal*, 10(3), 339–350.

Germanà, M.L. and Nicolini, E. (2022). Towards a sustainable use of architectural heritage: How to assess accessibility and the user's role. In *Conservation of Architectural Heritage. Advances in Science, Technology & Innovation*, Versaci, A., Bougdah, H., Akagawa, N., Cavalagli, N. (eds). Springer, Cham.

Guseva, L. (2018). Konstantin Dmitrievich Ushinsky: The founder of scientific pedagogy in the 19th century Russia. *History of Education & Children's Literature*, XIII(1), 479–491.

Hernández, G. (2021). Didactic principles: A proposal from the theory of subjectivity. *Culture & Psychology*, 27(4), 632–644.

Huizinga, J. (1982). *Homo Ludens*, translated by Zlata Tehova. Science and Art Publisher, Sofia.

Ildiz, G. and Ahmetoğlu, E. (2017). The Johann Heinrich Pestalozzi approach. In *Researches on Science and Art in 21st Century Turkey*, Arapgirlioglu, E., Atik, H., Elliott, A., Turgeon, R.L. (eds). Gece Kitapliği, Ankara.

Jůvová, A. and Bakker, F. (2015). Didactic principles by Comenius and 21st century skills. *e-Pedagogium*, 15(2), 7–20.

Khakimova, G., Azimova, M., Tuxsanova, V., Ibatova, N. (2021). Didactic principles in teaching fine arts. *Journal of Contemporary Issues in Business and Government*, 27(2), 5753–5759.

Krieg, S. (2011). The Australian early years learning framework: Learning what? *Contemporary Issues in Early Childhood*, 12(1), 46–55.

Leila, M.M.S.A. and El-Bastawisy, M.M. (2022). Rehabilitation and exploitation of heritage buildings. An investment approach. In *Conservation of Architectural Heritage. Advances in Science, Technology & Innovation*, Versaci, A., Bougdah, H., Akagawa, N., Cavalagli, N. (eds). Springer, Cham.

Marchigiani, E. (2022). The city as a learning context. Lessons of citizenship through the design of public spaces. In *Cultural Heritage Education in the Everyday Landscape. Digital Innovations in Architecture, Engineering and Construction*, Casonato, C. and Bonfantini, B. (eds). Springer, Cham.

Moskaliuk, E. (2018). Key educational concepts and students' worldview development elaborated by K.D. Ushinsky. *Science and Education a New Dimension. Pedagogy and Psychology*, VI(69), 165.

Steinrücke, J., Veldkamp, B., Jong, T. (2023). The effect of self-reflection on information usage and information literacy in a digital serious game. *Journal Computers and Education Open*, 4, 100133.

Vangsnes, V., Økland, N., Krumsvik, R. (2012). Computer games in pre-school settings: Didactical challenges when commercial educational computer games are implemented in kindergartens. *Journal Computers & Education*, 58(4), 1138–1148.

5

Educational Video Games on Climate Resilience of Built Cultural Heritage

In the last decades, many educational video games have been focused on the preservation of cultural heritage and, on the other hand, on climate extremes. However, no video games address preserving cultural heritage in the epoch of rapid climate changes. The proposed chapter is dedicated to presenting and comparing video games about climate resilience and the vulnerability of built cultural heritage regarding modern climate changes, together with some results about user experience received through their experimental validation.

5.1. Introduction

Since the introduction of the term serious games (SG) (Abt 1970), the meaning has evolved, but there is still more than one definition. However, most researchers agree that the primary purpose of SGs is related to some learning and education rather than only pure entertainment (Landers 2015; Susi et al. 2007; Michael and Chen 2005). Usually, game-based learning (GBL) is considered a subcategory of SG (Deterding et al. 2011; Kapp 2012). Educational video games (EVG) are a technological innovation that emerged due to the application of advanced ICT in the SG concept (Alvarez and Djaouti 2012). Such games can embrace various components, that is, learning content, game elements and digital objects representing essential aspects of the game context. These components are integrated into a well-composed virtual game plot design intended to meet the requirements of the target users. EVG provide students with high motivation and engagement while supporting enhanced learning (Martí-Parreño et al. 2016). Also, using laser projections (Ilchev et al. 2019) as innovative visualization technologies increase

Chapter written by Boyan BONTCHEV, Valentina TERZIEVA, Albena ANTONOVA, Yavor DANKOV and Dessislava VASSILEVA.

educational video game attractiveness and, thus, the involvement of students in the GBL process. Recently, such games have been increasingly integrated and widely used in various contexts for learning purposes. They can suit many social areas, for example, schools, universities, vocational education, healthcare, the military, fighting fake news and increasing cyber-security awareness (Laamarti et al. 2014). EVG enable students to acquire and test particular knowledge, make decisions and develop social skills (Connolly et al. 2014; Boyle et al. 2016). Moreover, EVG can teach skills beyond the classroom, for example, players can gain an experience in various virtual settings in the game context by making trials that could be impossible in the real world due to safety reasons or different constraints.

Climate change and its relation to the climate resilience of built cultural heritage are still new topics in education. However, SG and EVG can significantly influence education in this field (Wu and Lee 2015; Gerber et al. 2021). By integrating learning and entertainment, EVG are a powerful, persuasive tool that raises awareness and engagement, encourages attitude and behavior change (Gao et al. 2021), and stimulates decision-making (Marome et al. 2021). Climate change games typically have three primary objectives: teaching knowledge and providing topic acquaintance, raising awareness of the global warming challenges and encouraging solutions development (Flood et al. 2018). However, the potential of SG is broader than education or raising awareness (Chmutina et al. 2021). Well-designed EVG stimulate immersion and encourage critical thinking, collaboration, creativity and problem-solving skills, making them increasingly efficient in lifelong learning to reach a larger target audience. EVG, as a subset of SG, are effective in improving the understanding of complex issues and concepts. Their immersive nature allows the acquisition of skills and knowledge by playing, so when they are used as a teaching approach, they induce a positive change in behavior and attitude. Next, EVG provide multisensory interdisciplinary learning spaces that enable exploring and assessing various hypotheses to ensure a complete interpretation of the complicated issues of built heritage climate resilience (Bontchev et al. 2022). The ARCH project[1] develops standardized resilience-enhancing methods and tools to improve the resilience of historic built and cultural heritage (Lindner et al. 2021). The goal is to provide better information and decision-making support for the authorities about climate change and disaster risk-reduction measures. These tools are turned into a role-playing game to enhance resilience in historic building areas.

The chapter continues with section 5.2, presenting the classification of e-CREHA video games for climate resilience and built heritage. After that, section 5.3 introduces the APOGEE software platform for creating educational video maze games and the game design process. The results from two maze video games about

1 Available at: https://savingculturalheritage.eu/.

the resilience and vulnerability of built heritage are presented and discussed in section 5.4. The chapter ends with conclusions and contributions.

5.2. e-CREHA video games for climate resilience and built heritage

A short desktop study in the e-CREHA project[2] implementation scope revealed that climate resilience and built heritage protection topics are not addressed in publicly available video games. Since serious games are increasingly used to raise awareness in the scope of long-life-learning, project partners investigated available video games, covering some related topics. In addition, several papers were found, for reviewing the serious games for cultural heritage (Anderson et al. 2010), addressing how narrative-based games can help for teaching cultural heritage (Malegiannaki et al. 2020), or discussing the role of video games in architecture education (Elsamahy 2017). A summary of the desktop study findings was presented in the project outcome IO1 (Öngören et al. 2021).

A total of 37 video games were found, classified by several characteristics: game type, dimension (2D or 3D), game genre and online resources. All 37 video games are related to climate change or the cultural heritage field, but some of the video games referred to were unavailable (flash-based/expired/inaccessible host sites). The games are distributed as follows: 15 covered topics from the built cultural heritage, and 21 covered specific climate change issues. Most of the games are puzzle and action-adventure games. Concerning the e-CREHA project topics, the games mainly cover themes from resilience (27), impact (18), adaptation (13) and vulnerability (5). Most games are single-player (28), and only nine are multiplayer video games. Considering the graphic interface, most games (26) support simple 2D games, 10 games are a 3D visual interface and one is a 2.5D graphical interface. In conclusion, the desktop survey proved a strong potential for investigating serious games for promoting climate resilience in cultural heritage protection (Öngören et al. 2021).

5.3. APOGEE approach for generation of enhanced maze video games

The APOGEE software platform for creating educational video maze games provides opportunities for designing and developing games diverse in genre and content (Bontchev et al. 2019). The platform is presented in detail in several publications by the authors (Bontchev et al. 2020; Dankov et al. 2022). In general, the process of creating educational video mazes enhanced with didactic mini-games on the platform goes through three main stages: game design, game generation and game validation (Bontchev et al. 2019).

---

2 Available at: https://www.ecreha.org/.

The APOGEE platform provides opportunities for creating educational maze video games by users with different knowledge of information technologies. This is provided by various tools that support and assist all processes in the platform (Bontchev et al. 2020). This facilitates designing, creating and analyzing the games designed and played on the APOGEE platform. Each educational game is designed and developed for a specially defined group of users with the primary goal of meeting the requirements and specific characteristics of these users, engaging their attention, enhancing their user experience in the games and achieving the educational and pedagogical goals set in the game. For a game to be effective and teach its learners, designers must carefully design the game.

The designer starts designing the games on the platform, and the process goes through several main stages presented in Figure 5.1.

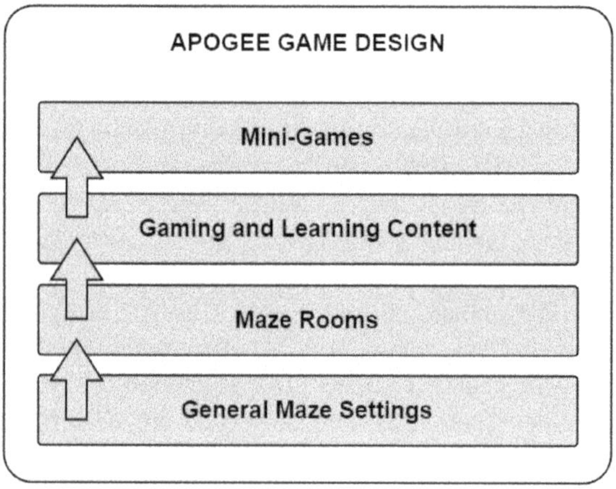

Figure 5.1. *APOGEE game design process*

The diagram represents the sequential steps the designer must go through and make the relevant decisions for the game to be generated and validated in the following stages. The approach to designing educational video games in the APOGEE platform is incremental and iterative, starting with defining the most general settings of the maze and then choosing the number of halls and the way of connecting between them. The designer must then select the educational and game content to be included in the game and the most appropriate mini-games in the respective maze hall. All information is documented in an XML document containing the formal game description. The XML document must comply with the predefined rules for creating an XML document or the so-called XML Scheme. The XML document, along with all selected educational and game content in the form of various game assets such as text, images, audio-video files, etc., is processed by a unique plug-in tool in Unity 3D called Maze Builder responsible for the validation and generation of the initial version of the game (Bontchev et al. 2019). Suppose everything is correct, and the designer is satisfied with the working version of the game. In this case, it moves onto the final generation of the maze-type educational video game players and learners can play.

5.4. Results from two maze video games about resilience and vulnerability of built heritage

The e-CREHA project proposal planned the construction and practical validation of four educational enhanced maze video games for raising awareness among the general audience and experts for resilience, vulnerability, impact and adaptability of built cultural heritage in the scope of climate extremes. At present, two of them have been built, namely, the resilience game "Let's save Venice" (Bontchev et al. 2022) and the vulnerability game ("Vulnerability of heritage of Molise region to landslide and erosion phenomena caused by/related to climate changes"). Both the games possess circular topology containing five halls. The halls represent educational content corresponding to their names: "Introduction", "The Context", "The Problem", "The Solution", and "The Future". The player's mission in the first maze game (about the resilience of built heritage) was to understand how Venice can be saved from ever-growing floods. In the second maze game, the mission was to explore the impact of climate variations on cultural heritage, its consequences and the main factors: exposure, resistance and resilience. To achieve the mission goals, the player must pass through all the maze halls, read the information on the learning boards, complete all mandatory tasks (presented by didactic mini-games) and find all hidden objects presenting solutions to the main problems. The resilience game contained 38 learning boards, six types of mini-games, and five hidden objects, while the vulnerability game had 40 learning boards, seven mini-game types and six hidden objects. The didactic mini-games enhancing the video maze were of seven kinds: (a) answering a question to unlock a door, (b) roll-a-ball, (c) finding hidden objects, (d) word soup, (e) memory game, (f) arranging pictures and (g) quiz (only for the vulnerability game). Screenshots of some of these mini-games are presented in Figure 5.2. Thus, the player should collect the maximum number of points (i.e. achieve the maximum result) by playing didactic mini-games in the shortest possible time. The playing time was planned to be 15–30 minutes, though some players found the game very interesting and impressive and played it for up to 1 hour.

The initial prototypes of the games were designed through the APOGEE platform for 3 months by following the design steps of the educational 3D video game maze (Bontchev et al. 2022). After receiving feedback from the project partners, they were enhanced and built as online educational video games.

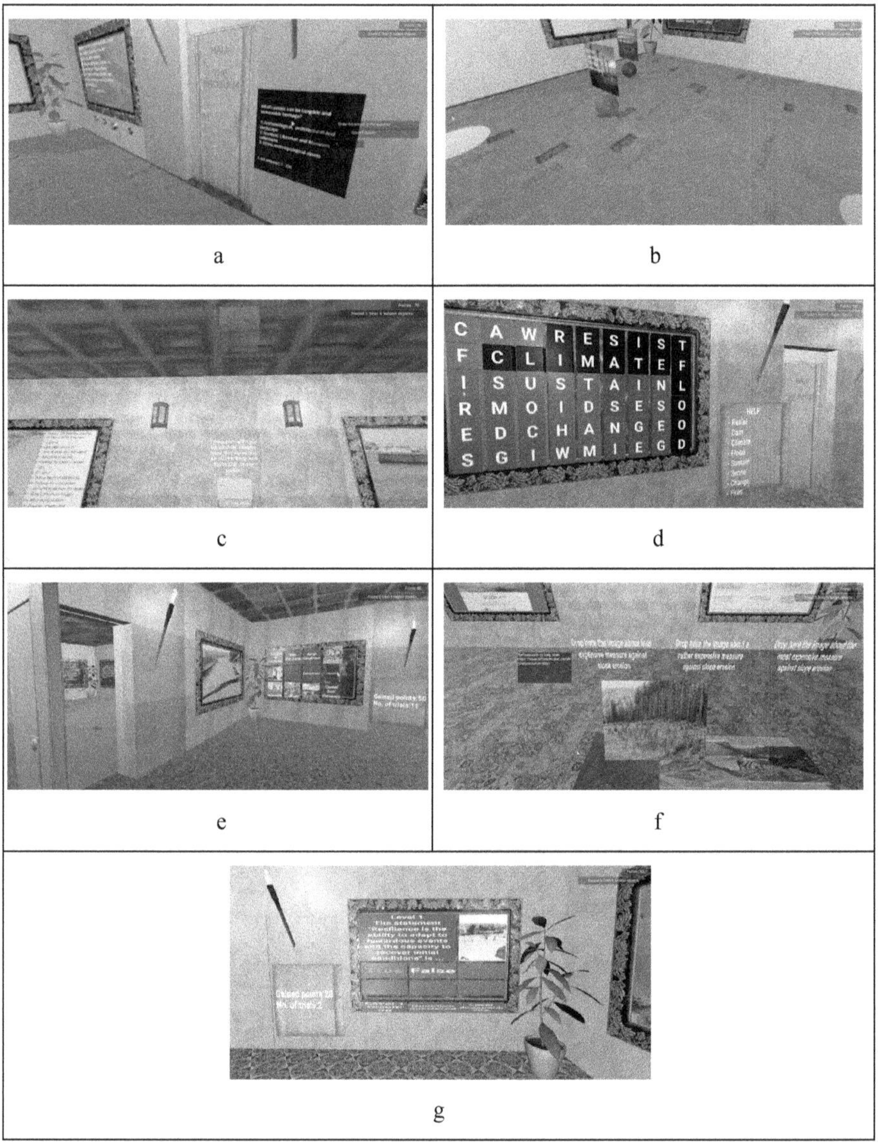

Figure 5.2. *Screenshots of didactic mini-games about resilience and vulnerability of built heritage. For a color version of this figure, see www.iste.co.uk/curulli/heritage.zip*

The practical validation of the resilience game took place at the first e-CREHA project workshop held from August 31 to September 8, 2021 at the Eindhoven University of Technology, in the Netherlands. The participants of the workshop were 21 graduate students and thee PhD students, with a mean age of about 25 years, from the e-CREHA partners' organizations: Eindhoven University of Technology (the Netherlands), TOBB University of Economics and Technology (Turkey), Sofia University "St Kl. Ohridski" (Bulgaria), and University of Molise (Italy). All 24 participants (12 males and 12 females) played the vulnerability game. The practical validation was conducted during the second e-CREHA project workshop held from September 13–22 2022 at the University of Molise (Termoli, Italy). In total, 22 players (six males/16 females, with a mean age of about 25 years) from the e-CREHA partners' organizations played the game.

For both games, students appeared to be inactive video game players – most reported playing video games less than 5 hours a week. Before the game session, students followed an introduction to the game, its learning and playing goals, and the rules of play. The gameplays were not limited in terms of time. On average, students played the resilience game for 23.52 minutes, with a mean number of found hidden objects of 3.75 (from 5) and 135.08 gained points. For the vulnerability game, the averaged values of playing time, hidden objects and accumulated points were 26.64 minutes, 5.364 (from six) hidden objects and 440.08 gained points (rather more because of the included quiz mini-game).

Game	Metric	Ease of learning	Familiarity	Consistency	Predictability	Informative feedback	Error handling
Resilience game	M	3.53571	3.89286	3.67857	3.85714	3.46429	3.39286
	SD	1.29048	1.16553	1.09048	0.93152	0.96156	1.03062
	SE	0.24388	0.22026	0.20608	0.17604	0.18172	0.19477
Vulnerability game	M	4.18182	3.95455	3.72727	3.63636	3.86364	3.77273
	SD	0.90692	0.72225	1.12045	0.84771	0.88884	0.86914
	SE	0.19336	0.15398	0.23888	0.18073	0.18950	0.18530

Table 5.1. *Attributes of learnability for both the resilience and vulnerability games*

After the gaming session, students did an online survey exploring their opinions about the maze game's learnability, playability and other characteristics. Table 5.1

presents mean values (M), standard deviation (SD) and standard error (SE) for six attributes of learnability for both the resilience and vulnerability games – ease of learning, familiarity, consistency, predictability, informative feedback and error handling. The five level Likert scale was applied (5 – very high, 4 – high, 3 – medium, 2 – low, 1 – very low). The vulnerability game demonstrated much better playability than the resilience game. Moreover, it motivated students to gain 3.26 times more points than in the resilience game. This can be attributed to enriched learning content and improved didactic mini-games embedded in maze halls.

5.5. Conclusion

Serious games are increasingly used to engage students' attention and achieve better academic performance. They are appealing because they are both motivating and educational. SG can provide an environment where students are immersed in an exciting environment with built-in learning objects, which improves their motivation, attitude and willingness to learn new knowledge. EVG can enable the implementation of user-centered pedagogical strategies and interactivity in the teaching process and conduct many practical simulations that are very expensive or dangerous in real-world conditions. They are included in the educational program of various fields, such as patient and health professional education, physics, chemistry, etc., where the possibility for practical demonstration and simulation is significant.

Developing serious games in a multidisciplinary field, such as climate resilience and the vulnerability of built cultural heritage to modern climate change, is a complex problem that requires the efforts of different types of professionals – architects, civil engineers, IT professionals and environmentalists. The APOGEE platform facilitates this process by allowing users with diverse knowledge of information technologies to create educational maze video games. This is provided by various tools that support and assist all processes in the platform.

The contribution of this research consists of the design, development and practical validation of two educational maze video games for raising awareness among the general public and experts for the resilience and vulnerability of built cultural heritage in the scope of climate extremes. The initial prototypes of the games are designed through the APOGEE platform by following the six design steps of the educational 3D video game maze. The practical validation was conducted by exploring the players' opinions about the maze game's learnability, playability and various IT characteristics. The students reported very high levels of ease of learning and familiarity, especially for the vulnerability game. Thus, the conducted experiments demonstrate the promising potential of enhanced maze games for acquiring competencies in climate change, climate resilience and protection of the built heritage. Hence, the research results presented will be applied to creating other educational video maze games for the impact and adaptability of built cultural heritage in the scope of extreme climate changes.

5.6. References

Abt, C.C. (1987). *Serious Games*. University Press of America, New York.

Alvarez, J. and Djaouti, D. (2012). *Serious Games: An Introduction*. Éditions Questions Théoriques, Paris.

Anderson, E.F., McLoughlin, L., Liarokapis, F., Peters, C., Petridis, P., de Freitas, S. (2010). Developing serious games for cultural heritage: A state-of-the-art review. *Virtual Reality*, 14(4), 255–275.

Bontchev, B., Vassileva, D., Dankov, Y. (2019). The APOGEE software platform for construction of rich maze video games for education. In *Proceedings of the 14th International Conference on Software Technologies (ICSOFT 2019)*, 491–498. SciTePress, Setúbal.

Bontchev, B., Antonova, A., Dankov, Y. (2020). Educational video game design using personalized learning scenarios. In *Computational Science and its Applications – ICCSA 2020*. Springer, Cham.

Bontchev, B., Antonova, A., Terzieva, V., Dankov, Y. (2022). "Let us save Venice" – An educational online maze game for climate resilience. *Sustainability*, 14(1), 7.

Boyle, E.A., Hainey, T., Connolly, T.M., Gray, G., Earp, J., Ott, M., Lim, T., Ninaus, M., Ribeiro, C., Pereira, J. (2016). An update to the systematic literature review of empirical evidence of the impacts and outcomes of computer games and serious games. *Computers & Education*, 94, 178–192.

Chmutina, K., Tandon, A., Kalkhitashvili, M., Tevzadze, M., Kobulia, I. (2021). Connecting heritage, vulnerabilities and capacities through a participatory game. *International Journal of Disaster Risk Reduction*, 53, 102005.

Connolly, T.M., Hainey, T., Boyle, E., Baxter, G., Moreno-Ger, P. (eds) (2014). *Psychology, Pedagogy, and Assessment in Serious Games*. IGI Global, Pennsylvania.

Dankov, Y., Antonova, A., Bontchev, B. (2022). Adopting user-centered design to identify assessment metrics for adaptive video games for education. In *Human Interaction, Emerging Technologies and Future Systems V*, Ahram, T. and Taiar, R. (eds). Springer, Cham.

Deterding, S., Khaled, R., Nacke, L.E., Dixon, D. (2011). Gamification: Toward a definition. In *Proceedings of the CHI 2011 Gamification Workshop Proceedings*, 12, 12–15.

Elsamahy, E. (2017). An investigation into using digital games-based learning in architecture education. *Architecture and Planning Journal*, 23(3), 19–30.

European Commission (2021). Gamifying tools to improve the resilience of Europe's historic areas to climate change [Online]. Available at: https://cordis.europa.eu/article/id/435389-gamifying-tools-to-improve-the-resilience-of-europe-s-historic-areas-to-climate-change [Accessed 10 January 2023].

Flood, S., Cradock-Henry, N.A., Blackett, P., Edwards, P. (2018). Adaptive and interactive climate futures: Systematic review of "serious games" for engagement and decision-making. *Environ. Res. Lett.*, 13, 063005.

Gao, W., Guo, Y., Jiang, F. (2021). Playing for a resilient future: A serious game designed to explore and understand the complexity of the interaction among climate change, disaster risk, and urban development. *Int. J. Environ. Res. Public Health*, 18, 8949.

Gerber, A., Ulrich, M., Wäger, F.X., Roca-Puigròs, M., Gonçalves, J.S., Wäger, P. (2021). Games on climate change: Identifying development potentials through advanced classification and game characteristics mapping. *Sustainability*, 13, 1997.

Ilchev, S., Petkov, D., Andreev, R., Ilcheva, Z. (2019). Smart compact laser system for animation projections. *Cybernetics and Information Technology*, 19, 137–153.

Kapp, K. (2012). *The Gamification of Learning and Instruction: Game-Based Methods and Strategies for Training and Education*. Pfeiffer, San Francisco.

Laamarti, F., Eid, M., El Saddik, A. (2014). An overview of serious games. *International Journal of Computer Games Technology*, 358152.

Landers, R. (2015). Developing a theory of gamified learning: Linking serious games and gamification of learning. *Simulation & Gaming*, 45(6), 752–768.

Lindner, R., Lückerath, D., Milde, K., Ullrich, O., Maresch, S., Peinhardt, K., Latinos, V., Hernantes, J., Jaca, C. (2021). The standardization process as a chance for conceptual refinement of a disaster risk management framework: The ARCH project. *Sustainability*, 13(21), 12276.

Malegiannaki, I., Daradoumis, T., Retalis, S. (2020). Teaching cultural heritage through a narrative-based game. *Journal on Computing and Cultural Heritage*, 13(4), 1–28.

Marome, W., Natakun, B., Archer, D. (2021). Examining the use of serious games for enhancing community resilience to climate risks in Thailand. *Sustainability*, 13, 4420.

Martí-Parreño, J., Sánchez-Mena, A., Aldás-Manzano, J. (2016). Teachers' intention to use educational video games: A technology acceptance model approach. *10th European Conference on Games Based Learning: ECGBL 2016*, 434–441. University of West Scotland, Paisley.

Michael, D.R. and Chen, S.L. (2005). *Serious Games: Games that Educate, Train, and Inform*. Thomson Course Technology, Boston.

Öngören, P., Acar, S., Özbek, A., Yılmaz, B., Curulli, I., Kaya, D., Guzman, P. (2021). O1-cross-disciplinary pedagogies. eCREHA project Intellectual Output 1 [Online]. Available at: https://www.ecreha.org/_files/ugd/48e8b5_7bc3a2f7c2da46d6b36f3a03eef14ac1.pdf [Accessed 10 January 2023].

Susi, T., Johannesson, M., Backlund, P. (2007). Serious games – An overview. Technical report HS-IKI-TR- 07-001, School of Humanities and Informatics, University of Skövde.

Wu, J. and Lee, J. (2015). Climate change games as tools for education and engagement. *Nature Climate Change*, 5, 413–418.

6

Heritage Education for Climate Justice and Adaptation

The roles of heritage conservation in sustainable development and climate mitigation are increasingly recognized in teachings and research that bridge these areas. Climate impacts and the need for adaptation are already challenging the established focus areas of heritage conservation education. Furthermore, decolonization and anti-racism compel us to question the heritage we conserve and for whom. How can we adapt and transition education in heritage and conservation to address the growing impacts of climate change and the need for more diverse voices to contribute in setting priorities for adaptation? This chapter reports on an education initiative entitled *Just Transitions: Heritage Education for Climate Adaptation* (JTHECA), which responds to this question.

6.1. *Just Transitions: Heritage Education for Climate Adaptation*

Climate change, like heritage, is a subject that both requires and generates interdisciplinary approaches. The Climate Heritage Network (CHN) is a sweeping network of arts, cultural and heritage organizations focused on tackling climate change. The CHN includes academic departments from across the arts, humanities, sciences and engineering. Started in 2021, JTHECA is an initiative of academic institutional members of the CHN working in Canada, notably at Carleton University in Ontario, Athabasca University in Alberta, and Université de Montréal in Quebec. Our objectives are to bring together educators in heritage conservation, to better define the contributions our field can bring to climate issues, including

---

Chapter written by Susan ROSS.

mitigation, adaptation and transitions toward climate justice. A key concern is to be inclusive of all voices and perspectives. In the Canadian context, there is a growing recognition that Indigenous heritage, knowledge and governance have an important place in these discussions[1].

JTHECA's initiatives build on existing disciplines and interdisciplinary approaches from architecture, engineering, history heritage planning and studies, while looking to identify expertise from other areas – such as geography, environmental sciences and humanities, biodiversity sciences and Indigenous relations with the land – that is needed to define, develop and deliver new approaches, knowledge areas and skills. Parallel objectives include understanding how to effectively communicate the key messages about the role of heritage conservation in climate action to emerging scholars and practitioners and emphasizing the role of universities in identifying the methods, subjects and partnerships that should be prioritized in teaching and research.

JTHECA's initial plan was to host events that would make a difference to academics – including faculty and students – who are grappling with how to keep heritage conservation relevant and effective in these times of change. We began a series of three events in June 2022. The first event – directed at educators themselves – focused on identifying and communicating the knowledge areas to be developed in education for climate change. Invited experts discussed topics that will alter what and how we should teach, including climate risk assessment, building resilience and adaptation planning, resource equity, material reuse and trades, Indigenous heritage and governance, and climate justice. Given the importance of the Conference of the Parties (COP) summits under the United Nations Framework Convention on Climate Change (UNFCCC) to the broader CHN, the February 2023 event responded to and reflected upon the inclusion of cultural heritage at COP 27. Invited speakers who had attended COP 27 on behalf of advocacy, Indigenous and government organizations shared their perspectives in "Cultural Voices from Arts and Heritage at COP 27" (JTHECA). Subsequent events will showcase new educational models and involve students in learning through case studies.

A wide range of scholarship and pedagogical models are now available to inform how we advance the efforts of heritage education as part of just transitions and climate adaptation. To support the event discussions, a website was created to share resources, including recordings of JTHECA events (JTHECA).

1 The organizers are also all active members of the National Roundtable on Heritage Education: an organization that gathers educators across all heritage sectors from public education to university degree programs.

6.2. ICOMOS University Forum – June 3–4, 2022

The first event of JTHECA was organized in partnership with ICOMOS/ ICOMOS Canada as the first online ICOMOS University Forum[2]. ICOMOS (the International Council on Monuments and Sites) is "a global non-governmental organization associated with UNESCO, whose mission is to promote the conservation, protection, use and enhancement of monuments, building complexes and sites". The overarching goal of the ICOMOS University Forum "is to launch a flexible operational forum where global cooperation among universities and ICOMOS can quickly lead to identifying pressing issues and to incubate and develop creative solutions". This Forum focused on identifying and defining knowledge areas for heritage education to support adaptation of places and communities to changing climates. According to the Adaptation Learning Network's Climate Adaptation Competency Framework:

> Climate adaptation requires a foundation of knowledge or literacy in a number of scientific areas and from a number of different worldviews and perspectives. These include understanding and being able to consider, bridge, and apply knowledge drawn from Western climate change science and climate models, and Indigenous knowledge systems. It requires understanding and being able to apply systems thinking and climate adaptation science to a range of issues and opportunities (Cox et al. 2020, p.17).

Over the years, ICOMOS has produced several reports that help identify specific competency areas to be developed in heritage practices. This event's discussion recognized the value of many of the recent ICOMOS reports on climate change as foundational resources. Notably, the 2019 ICOMOS report *Future of Our Pasts: Engaging Cultural Heritage in Climate Action* provides specific recommendations related to education and adaptation, including calling for:

> Training and education for the implementation and monitoring of adaptation actions, disaster planning and recovery. Training and capacity building to ensure the correct skills, materials and procedures (e.g., emergency evacuation of movable heritage) including inter-sectoral and inter-disciplinary cooperation. Training in recovery strategies could include heritage-based processes to foster social cohesion (ICOMOS 2019, article 7.6).

2 The ICOMOS University Forum/JTHECA planning team consisted of Shabnam Inanloo Dailoo, Claudine Déom, Abi Kang, Susan Ross, Mario Santana Quintero and Laurie Smith. I would like to acknowledge the assistance of Laurie Smith in reviewing this chapter.

> Harnessing the power of heritage to create public engagement/education on climate change, creating dialogues within and between communities (ICOMOS 2019, article 10.2.1).

While many of the education-related recommendations in *Future of Our Pasts* suggest the importance of knowledge exchange across sectors and generations, there are specific opportunities in academic and educational contexts to focus on critical pedagogical development, to engage with diverse communities, to make place for Indigenous knowledge and land-based learning and to foreground adaptation strategies for the campus properties (Maina-Okori et al. 2018; Jorgenson et al. 2019; Pereira et al. 2021; Johnson et al. 2022). Recognizing the specific challenges of university-level education, the overarching objectives of this ICOMOS University Forum included:

– Help for developing a shared understanding of basic knowledge gaps or strengths, by offering an introduction and overview of the areas of study to be included in heritage education that will support increased competency in graduating students.

– Enabling educators to hear from experienced peers about opportunities for and barriers to integration of these knowledge areas in existing heritage education in diverse disciplines, professional development and training contexts.

– Equipping participants with an overview of available resources in these areas, ranging from academic literature to pedagogical tools and other resources.

The program included four themed discussion panels spread over 2 days: "Climate Impacts and Heritage Vulnerability"; "Building/Urban Adaptation and Resilience"; "Materials/Resources, Reuse and Equity"; and "Climate Justice, Indigenous Knowledge, and Governance". We identified themes by surveying academic members of the Climate Heritage Network; selected themes represented the priority areas of the organizers and the speakers. Review articles that systematically address the relationship between climate change and (cultural) heritage (conservation) also help to identify gaps (e.g. Fatoric and Seekamp 2017; Orr et al. 2021). We also recognized the limitations of our knowledge areas and experiences in these emerging areas as Canadian academics.

The goal of each panel was to identify and discuss key concepts to support developing competencies in climate adaptation in heritage education. Subject area experts and experienced educators in identified sub-areas discussed the main concepts that are central to understanding these areas, as well as aspects of applicable educational approaches to be considered.

6.3. Knowledge areas for heritage education: panel theme overviews

To facilitate the entry into the Forum's discussions, each day began with a presentation on a major concept of JTHECA focus: Thea Dickinson (University of Toronto/Environment Canada/IPCC) spoke on climate adaptation on June 3, and Georgina Cundhill-Kemp (IDRC, Rhodes University) spoke on climate justice on June 4. These speakers working outside heritage conservation were invited in to help situate the questions of JTHECA in the larger themes of climate adaptation and climate justice. ICOMOS's 2019 report *The Future of Our Pasts, Engaging Cultural Heritage in Climate Action* provides these two definitions:

– *Adaptation (in human systems)*: "The process of adjustment to actual expected climate and its effects, in order to moderate harm or exploit beneficial opportunities. In natural systems, the process of adjustment to actual climate and its effects; human intervention may facilitate adjustment to expected climate and its effects" (p. 92).

– *Climate justice*: "Climate justice links Human Rights and development to achieve a human-centered approach to climate action, safeguarding the rights of the most vulnerable; taking into account the needs of those at the greatest risk, particularly the poorest and most vulnerable; and sharing the burdens and benefits of climate change and their resolutions in an equitable and just manner" (p. 100).

Each panelist started by speaking briefly about their knowledge area and the potential educational context, format or needs to consider. This was followed by a facilitated panel discussion, then an open discussion taking questions from participants.

Climate impacts and heritage vulnerability

To begin to plan for adaptation, we need to understand impacts and how to assess them. This will allow us to consider the relative vulnerability of heritage places to critical changes, both those which may come about slowly, or increasingly, those that arrive through more drastic events. In this session, we invited panelists to share their expertise and experiences with concepts that included climate impacts and vulnerability assessment, teaching climate change in communities, and teaching and planning disaster risk preparedness and reduction. Speakers on this panel, moderated by Mario Santana (ICOMOS/Carleton University/CHN), included Cathy Daly (Lincoln University/CHN), Will Megarry (Queens University/ICOMOS/CHN) and Rohit Jigyasu (ICCROM/CHN).

Building/urban adaptation and resilience

Much of the focus on heritage conservation for the built environment, including for historic urban areas, has been around climate mitigation. It is important to examine the relationship between mitigation and adaptation to help see the additional areas that we will need to understand in the context of adaptation. In this session, we invited panelists to share their expertise and experiences with concepts that included building resilience and adaptation, planning adaptation for the Historic Urban Landscape and teaching climate mitigation and adaptation in built heritage. Speakers on this panel, moderated by Claudine Déom (Université de Montréal/ CHN), included Oriel Prizeman (Cardiff University/APT), Paloma Guzman (Norsk Institut KU/CHN) and Mark Brandt (MTBA/ZNCC/APT/CHN).

Materials/resources, reuse and equity

Reuse of existing materials and resources is at the heart of heritage practice. Strategies for both climate mitigation and adaptation draw attention to the wider built legacies of human cultures, including the impact of extractive industries that are embodied in the building materials being reused during adaptation. Reuse in the context of social justice must address equitable access to resources. In this session, we invited panelists to share their expertise and experiences with concepts that included the role of adaptive reuse and circularity within the Historic Urban Landscape approach, methods for assessing embodied effects of existing buildings and the roles of traditional and sustainable building materials. Speakers on this panel, moderated by Susan Ross (Carleton University/CHN), included Deniz Ikiz Kaya (TU Eindhoven/CHN) and Lori Ferris (Goody Clancy/ZNCC/APT/CHN).

Social justice, Indigenous knowledge and governance

Emerging concepts that bridge topics of social justice and climate adaptation include "just transitions" and "building back better". Furthermore, Indigenous knowledge systems and governance models suggest alternate strategies for building healthy relationships between communities of human and nonhumankinds and the planet. In this session, we invited panelists to share their expertise and experiences with concepts that included the relation of social justice to cultural heritage, Indigenous heritage, and community planning, and teaching Indigenous Environment/Land Issues. Speakers on this panel, moderated by Shabnam Inanloo Dailoo (Athabasca University/CHN), included Lorna Crowshoe (City of Calgary, National Trust for Canada) and Courtney Vaughan (Archipel Research, Carleton University).

Student and emerging professional views

To conclude the educator-focused initial event, at the end of each day, a student and an emerging professional were invited to reflect on how knowledge areas discussed build upon and or challenge what they have been recently learning or experiencing in their practices. The student comments amplified the observation of existing gaps but also how general skills such as documentation will remain critical, and how progress in transforming education is already underway. Emerging professionals in ICOMOS are already also organizing events like webinars to address climate change questions. Shifts in conservation education that address the questions students have about climate, sustainability and the relation of academic theories to practice are already happening. This includes expanding from the building to the urban scale, to include communities, to consider processes and systems, and connect intangible and tangible heritage.

Discussion

Importantly, these four panels helped to remind us of strengths in knowledge areas while identifying gaps. In the heritage context, improving existing practices is valued, but we recognize increasingly that we are at the cross-roads of shifting values. This can however increase the opportunities to transform critical relations that have been defined too distinctly. We have seen that young people represent a powerful force, and we need to use iterative, reflective and reactive learning to support new roles. We will need to expand on how we use our connections to communities to make essential links and support greater responsibility in our actions, while being humble about what we need to learn from those who have often been left out of our consideration of heritage and climate.

6.4. Land acknowledgement(s) and climate heritage education

JTHECA aspires to address much-needed capacity development for decolonizing education in heritage and climate adaptation. This will include ensuring that the knowledge we are seeking to define and develop is informed by Indigenous knowledge, including helping to reframe Western science and its contributions to the causes of climate change.

The land acknowledgement is an Indigenous protocol that positions the speaker in relation to the land, recognizes the Indigenous rights-holders of the land and names specific First Nations, Métis and Inuit communities associated with the land.

It has become an important element of Canadian university events. The land acknowledgement used at JTHECA events and posted on their website provides a departure point for ongoing learning and exchange in the context of Indigenous knowledge.

"*Pîjashig! Kwe kwe! Tunngahugit! She:kon! Aanii! Boozhoo! Tansi! Taanishe! Hello! Bienvenue!*" Greetings in multiple languages of the First Nations, Métis and Inuit peoples on the JTHECA website honour the fact that these events are being organized by members of academic institutions located across Canada, with participants from other parts of Turtle Island (North America) and the world beyond. We wish to acknowledge the Indigenous peoples who are our hosts, and our ongoing responsibilities to all forms of life in these lands and waters.

Specific land acknowledgements are made for each event by lead institutions and by individual presenters at their own initiative[3]. The importance of acknowledging the land is amplified by our roles as educators in these territories, where we can help develop awareness of Indigenous cultures, to engage with Indigenous leaders and communities in how we teach, and to discuss how to bring Indigenous knowledge into teaching, including Indigenous ideas of governance (Vowell 2016). Through this, we can learn how to provide critical alternatives to the exploitation of the lands, waters and all life forms, which has so long profited settlers at the expense of the lives, health and culture of Indigenous peoples. We recognize that the project of decolonizing education is even more important following the Truth and Reconciliation Commission of Canada's *Calls to Action* which identified specific roles for educational institutions, and education in every aspect of society (Truth and Reconciliation Commission (TRC) of Canada 2015).

6.5. Outputs and dissemination

An important part of developing capacity in heritage education is to share the resources compiled in identifying themes and panelists for JTHECA events. The June 2022 and February 2023 events were recorded, and these recordings are available on the JTHECA website. The website also includes a section on Resources organized by major theme areas: Climate Impacts, Climate Adaptation, Education, Indigenous Knowledge and Governance, Resources/Materials and Tourism (JTHECA). Future additions will include teaching resources, including videos of related online webinars and conferences, and course syllabi and textbooks.

3 To identify the Indigenous territory where you are located, see Native Land Digital (n.d.).

6.6. References

Cox, R., Niederer, S., Forssman, V., Sikorski, L. (2020). Climate adaptation competency framework. The resilience by design lab: Adaptation learning network [Online]. Available at: https://can-adapt.ca/.

Fatorić, S. and Seekamp, E. (2017). Are cultural heritage and resources threatened by climate change? A systematic literature review. *Climatic Change*, 142(1–2), 227–254.

ICOMOS. (n.d.). ICOMOS University Forum [Online]. Available at: https://www.icomos.org/en/what-we-do/disseminating-knowledge/icomos-university-forum.

ICOMOS (2019). The future of our pasts: Engaging cultural heritage in climate action [Online]. Available at: https://civvih.icomos.org/wp-content/uploads/Future-of-Our-Pasts-Report-min.pdf.

Johnson, D.E., Parsons, M., Fisher, K. (2022). Indigenous climate change adaptation: New directions for emerging scholarship. *EPE Nature and Space*, 5(3), 1541–1578.

Jorgenson, S.N., Stephens, J.C., White, B. (2019). Environmental education in transition: A critical review of recent research on climate change and energy education. *The Journal of Environmental Education*, 50(3), 160–171.

Just Transitions: Heritage Education for Climate Adaptation (JTHECA) (n.d.). [Online]. Available at: https://carleton.ca/jtheca/.

Maina-Okori, N.M., Koushik, J.R., Wilson, A. (2018). Reimagining intersectionality in environmental and sustainability education: A critical literature review. *The Journal of Environmental Education*, 49(4), 286–296.

Native Land Digital (n.d.). [Online]. Available at: https://native-land.ca.

Orr, S.A., Richards, J., Fatorić, S. (2021). Climate change and cultural heritage: A systematic literature review (2016–2020). *The Historic Environment: Policy & Practice*, 12(3–4), 434–477.

Pereira, L.D., Tavares, V., Soares, N. (2021). Up-to-date challenges for the conservation, rehabilitation and energy retrofitting of higher education cultural heritage buildings. *Sustainability*, 13(4), 1–12.

Truth and Reconciliation Commission (TRC) of Canada (2015). TRC calls to action [Online]. Available at: https://ehprnh2mwo3.exactdn.com/wp-content/uploads/2021/01/Calls_to_Action_English2.pdf.

Vowell, C. (2016). Beyond territorial acknowledgements. Apihtawikosisan [Online]. Available at: https://apihtawikosisan.com/2016/09/beyond-territorial-acknowledgments/.

7

Education for Climate Resilient Architectural Heritage (e-CREHA): Multidisciplinary Blended Learning for Climate Resilient Heritage

7.1. Introduction

The recently published Global Research and Action Agenda on Culture, Heritage and Climate Change that is a key outcome of the ICOMOS, UNESCO and IPCC International co-sponsored meeting on Culture, Heritage and Climate Change, states that "heritage and cultural practices can act as a bridge between different ways of knowing, embody inherited knowledge accumulated over generations, and serve as entry points for innovative and inclusive solutions to climate action" (ICSM CHC 2022, p. 2). This requires acknowledging the role of culture and heritage in transformative change and alternative sustainable futures. The loss or deterioration of architectural heritage can have a negative impact on national and local communities, due to both its cultural importance as a source of information on the past and local identity, and its socioeconomic value. This indicates that heritage contributes to social cohesion, sustainable development and psychological well-being. Therefore, by protecting heritage we promote resilience.

Architecture discipline has a key role in achieving both resilience and sustainable development of Europe to comply with the objectives of the Paris Agreement and the EU 2050 climate goals (European Council 2014). In fact, architecture is a key

---

Chapter written by Irene G. CURULLI, Deniz IKIZ KAYA, Rossella NOCERA and Luciano DE BONIS.

profession in reducing the effects of climate change, for example, energy use in buildings is one of the major producers of greenhouse gases. Accordingly, awareness and knowledge of the role that the built environment plays in the climate crisis are triggering demands for future architects' skills in sustainable and resilient design. The Architects' Council of Europe held at the European Parliament in 2012 explicitly highlighted the architects' role in the process of adaptation to climate change. Therefore, a robust environmental education must sit at the core of the architectural curriculum in both academic and professional fields (Architects' Council of Europe (ACE/CAE) 2012).

Education and the problem of climate change in architectural design and planning have become increasingly topical during the last decennia; furthermore, there is an ongoing debate on how the possible methods of architectural education could merge theoretical and practical knowledge when facing climate change (Stupar et al., 2017). In this context, the curriculum has given very little attention to (or neglected) the vulnerability of the built heritage to climate change and has overlooked the contribution of heritage to the development of climate-adaptive strategies for a resilient society (Brabec et al. 2015; Holtorf 2018; ICOMOS 2019). Furthermore, this deficiency has pointed out two important needs: first, to introduce innovative learning methods to ease the understanding of the values and characteristics of built heritage; second, to adopt an interdisciplinary approach to education as the most appropriate framework for studying the complexity of climate change (Mochizuki et al. 2015) and to overcome the current divide between research and education. To achieve these goals, education should adopt methods that create a high level of interactivity and adjustability to changing environments, directed toward critical thinking (Šuligoj et al. 2020; Gonzales-Cacho et al. 2022). In short, education enables a new set of skills to be built that allow a comprehensive climate-sensitive and environmentally friendly professional action.

This chapter addresses the design, programming, supervision and evaluation processes of the e-learning course entitled "Education for Climate Resilient European Architectural Heritage" (e-CREHA), developed as part of the relevant project funded by the European Union Erasmus+ KA203 program. This three-year project was developed in collaboration with the Eindhoven University of Technology (coordinator), Institut National des Sciences Appliques Strasbourg, TOBB ETU, Sofia University, University of Molise, NIKU Research Institute and Beeldland.

After a brief introduction of the course's aim, methodology and structure, we will provide a description of the course modules and the analysis of the evaluation activities that were adopted to validate the course.

7.2. Objectives and methodology

The aim of the e-CREHA e-learning course is to develop multidisciplinary knowledge for building a culture of prevention and mitigation/adaptation, or a culture of preparedness, that recognizes the uniqueness of architectural heritage and its contribution to the development of climate actions. From this perspective, this e-learning course promotes the synergy between research and education and has a dual aim: to increase our understanding of how heritage and design can strengthen society's awareness of climate change, and to show how heritage and design play an effective role in shaping vulnerable environments while preserving them. Accordingly, the relationship between built heritage and resilience can be understood as a result of negotiation through which some past is selected for protection and representation of the present and to leave a message for the future (Graham and Howard 2008).

This online course is built on the collaborative exchange of multiple disciplines and expertise including architecture and built environment, heritage studies, engineering, design, climate science and software technologies and informatics. In this way, it expands students' intellectual resources, tackles the skill gaps necessary to manage change and provides them with a new climate-sensitive set of skills to build resilience for European cultural heritage.

Methodology and structure of the course

The e-CREHA online course adopts a blended learning (BL) methodology. BL converges two archetypical learning environments: the traditional face-to-face learning environment and computer-mediated instruction. The latter is the result of the rapid emergence of technological innovations/digital technologies that have expanded the possibilities for distributed communication and interaction, with a huge impact on the opportunities for learning environments (Graham 2022).

The blended design of the e-CREHA course strategically combines online and in-person learning, and it can be thought of as having two interwoven layers: the physical (structural) layer and the pedagogical one (Curtis et al. 2006). The physical layer provides the rhythm of switching between in person-activities and online activities and the use of a particular learning platform for online activities. The pedagogical layer consists of teacher lectures, online groups and content interactions that are part of the active learning strategies. This layer directly influences student learning.

The e-learning course provides tailored training to masters level higher education students from multiple backgrounds. The course is delivered in four sections focusing on:

a) *Resilience* – built heritage and resilient concepts. The focus is the intersection of cultural heritage, climate change and resilience, the climate change risks for built heritage and the comprehension of sources of climate data and weather forecasts.

b) *Impact* – built heritage and impact of climate change. It focuses on climate heterotopia and climate-driven modifications of heritage by identifying the materiality and energy aspects of architectural heritage, and the design, transition and transformation aspects of heritage architecture.

c) *Vulnerability* – cultural heritage as a resource for climate change. It focuses on the concepts of vulnerability, fragility, antifragility and spatial data analysis and modeling (GIS) for managing the vulnerability of natural and built heritage to climate change.

d) *Adaptation* – resilient heritage solutions. It focuses on heritage and climate policies, knowledge tools and methods, and it presents best-practice approaches for climate adaptation.

Each section is organized into modules consisting of several integrated activities, such as teachers' thematic lectures, video clips and video games, interviews with experts, case studies, assignments and students' virtual forum(s). Each section ends with a final essay focusing on a heritage site as a case study, requiring students to carry out fieldwork and provide a discussion of their findings through the e-CREHA platform. Study cases vary in scale, geographic location and climate conditions. In educational terms, study cases contribute to the development of critical thinking and effective problem-solving skills as they initiate a process of participatory learning that promotes active and reflective learning. Accordingly, this approach nurtures the development of self-directed lifelong learners (Tomey 2003).

The modules are built on the following principles: (a) strategies of integration of different disciplines; (b) key issues of climate change and resilience, also embracing cultural, social and economic aspects; (c) prospects to capture participation from a large variety of students from multi-disciplinary backgrounds (diverse culture, knowledge background, practice); (d) prospects for building a new climate-sensitive set of skills, clear and up-to-date communication techniques, theoretical and practical knowledge from different disciplines; (e) critical analysis and strategies for the recognition of heritage as a catalyst and driver for sustainable development in students' future professional actions. The maze video game and video clips combine with modules to increase the impact on both the effectiveness and efficiency of learning. Literature studies prove that video and video games are very useful in developing skills in problem-solving, decision-making, strategy planning and assessment, while encouraging the growth of spatial representation skills (de Aguilera and Mendiz 2003). In particular, the pedagogical design of the videos is based on the “Three I’s Framework” – Image, Interactivity and Integration (f) (Asensio and Young 2002), and focuses on the key concepts and problems of the four sections of the course, and also suggests ad hoc research done by the students in order to provide possible solutions.

A customized Moodle platform is used to host the online course. Moodle (Modular Object Oriented Dynamic Learning) is a free open-source software package designed to develop learning platforms that are easy to design and used by both educators and students. It is a learning management system (LMS) that offers a huge variety of features such as course management, content implementation and delivery, collaboration tools, assessment, grading and reporting. The main reason for choosing the Moodle platform is its availability as a customizable, user-friendly and open-source platform. Its intuitive interface has been further adapted and customised to the purpose of the project. Its architecture is built according to the course sections and is conceived as a “community” platform that eases access to other universities interested in integrating the full/part of the course in their own curriculum. Each section of the course includes a set of lessons, assignments, tests/quizzes and interaction among students and teachers. After registration, it is very easy to access the platform, navigate through the sections of the course and follow its modules. The platform has been developed with the aim to increase students’ autonomy, motivate collaboration and encourage inclusion.

7.3. Course assessment activities

To assess the consistency of the goals and objectives of the e-CREHA course and the effectiveness of students’ learning, three activities were developed: (a) intern review of each section’s goals in addressing the existing gap in architectural education on climate-responsive cultural heritage; (b) development of three Intensive Study Programs’ (ISPs) design workshops to apply course principles, test the transfer of knowledge and validate the acquisition of new skills by the students; (c) taking corrective actions.

This section discusses the methods and results of the assessment of two sections of the online course, namely resilience and vulnerability, that were carried out through two ISPs-design workshops. The first one, ISP1-Climate Xtremes and Resilient Heritage, was held at the Eindhoven University of Technology (The Netherlands) in 2021 and the second one, ISP2-Climate Impacts and Heritage Vulnerabilities, was held at The University of Molise (Italy) in 2022.

The formula of ISPs-international design workshops is well known in architectural education as a timely platform/framework for learning. Its aim is to develop innovative knowledge and training concepts, promote intercultural dialogue (UNESCO 2009) and cross-disciplinary knowledge, integrate research with innovative teaching methods, and connect local and global needs through research and design. In particular, both ISPs adopted the research-by-design methodology. Accordingly, design is explored as a method of inquiry by the development of a project and also by exploring the different materials by which a design is carried out – sketches and mapping, among others (De Queiroz Barbosa et al. 2014). Critical questioning is central in the process of enquiry and it is assessed through interrelated and iterative actions (define, analyze, test, assess, re-test, re-assess). During the ISPs-design workshops, both tutors and students conducted research, tested and shared knowledge through interaction by means of learning-by-doing. This means that the organizational structure of both workshops intermingled theoretical and practical activities that focused on specific heritage sites and gave the opportunity to students and teachers to experience the local and broader significance of heritage directly, to assess its values critically and develop innovative design ideas in response to specific conditions of vulnerabilities and adaptive capacity with regard to climate change. In short, the ISPs-design workshops simulated the pedagogical layer of the BL method adopted by the e-CREHA course and used diverse pedagogies to explore climate resilient heritage issues.

7.4. ISPs-design workshops 1 and 2

Both workshops were organized according to the three overlapping phases of the research by design methodology, namely pre-design, design, and post-design phases. The active learning strategies consisted of site visits, fieldwork, studio-based work, presentations by experts and students, and meetings with practitioners and local communities.

About 30 students and 12 tutors from a range of disciplines, such as architecture and planning, geography, heritage and tourism studies, and computer sciences, participated in the workshop. Cross-disciplinary viewing of heritage and integration of knowledge through an interdisciplinary approach were priorities in the workshops' activities. The aim of the first workshop was to create a resilient landscape that is able to deal with unexpected floods. Therefore, the case studies focused on three sites of the Dutch watermill landscape in the North-Brabant region (The Netherlands) that are threatened by seasonal floods and changes in land use. The objective in the second workshop was similarly challenging: the students needed to assess heritage vulnerability and develop future adaptation scenarios for four given study cases located in Molise (Italy). The sites of archaeological, architectural and landscape value are threatened by fires, floods and landslides potentially caused by climate change.

Pre-design

At the start of the workshops, the video modules and thematic videos were used to introduce the specific topics of each ISP and were combined with lectures by local experts and academics. Through "open discussion sessions" and "group discussions" students debated the content of the modules in relation to the lectures and learned about the different problems and opportunities given by each specific site. During this pre-design phase, all students and tutors visited the proposed spatial areas, met with local authorities and interacted with local inhabitants.

The second step was to collect, map and assess information regarding the heritage qualities and historic characteristics of the selected area of investigation and relate them to the current policies and expectations. Besides the regular mapping on elevation, topography, water, land-use and urbanization in the Dutch watermill case studies (ISP1), maps were produced to give spatial insights into the status of river protection, climate data and risk, existing qualities of the historic landscape and to understand the landscape evolution over time. In a similar way, the second step of the pre-design phase of the ISP2 consisted of creating a georeferenced database regarding the immovable cultural heritage (archaeological, architectural and landscape), the hazards of fire, flood and landslide caused by climatic impacts, and finally, the land use patterns of the selected sites.

Students applied GIS tools to map, model and analyze the spatial data of the four case studies, benefitting from the instructive videos provided by the "vulnerability" module on GIS and web mapping tools. Thus, students acquired new skills for the management of heritage vulnerability to climate change.

Structured as a seminar course, the pre-design phase opened up the debate on the different cultural heritage concepts and for all the participants, it represented a stage for reflection on how heritage should be defined to become a useful resource for a climate-resilient landscape. Students started crafting problem statements and framing design goals to be achieved through the design proposals (Figure 7.1).

Figure 7.1. *Students engaged in discussion during the pre-design phase. For a color version of this figure, see www.iste.co.uk/curulli/heritage.zip*

Design phase

The design phase was the heart of both ISPs. Students developed design proposals that were the outcome of intense interdisciplinary knowledge exchange. Interaction between students, stakeholders and experts intensified. The multicultural and cross-disciplinary environment provided an excellent forum for speculative ideas on climate-resilient heritage. During studio work, students explored and synthesized multiple perspectives: climate forecast analysis combined with technical investigations (water engineering, fire risks, etc.), while developing sustainable tourism strategies or dealing with local heritage policies or sociocultural explorations of communities, etc. In both workshops, all projects embraced time and natural processes as serious factors and developed environments that have the capability to change and accommodate uncertainties. The video games on resilience and vulnerability helped students to assess what they learned and view the issues with a more critical eye while testing their research skills and acquisition of theoretical knowledge. All proposals were critically questioned on the meaning and impacts of their design propositions for the problems, as well as the consequences of the ideas for the potential qualities and opportunities for the area. After this, a process of continuous design, review and questioning, and redesign took place, during which new questions and objections were proposed and the design propositions were enriched.

In the "Resilient" workshop held in Eindhoven in September 2021, the outputs were four different design strategies that enhanced the synergy between heritage conservation and adaptation of the historic watermill landscapes, where they also acknowledged the social and cultural values of water. For instance, a proposal integrated water engineering with recreational tourism and flood prevention. It introduced a hydraulic pumping system that reinterpreted the water dynamics of the old watermill landscape in order to reveal and experience the historic landscape layers, re/connect the watermill landscape by disclosing the old paths and designing a net of new ones, preserve and value specific locations as secret places to be discovered and mitigate climate extremes with a system of water reservoirs for water collection and cooling (Figures 7.2(a) and (b)).

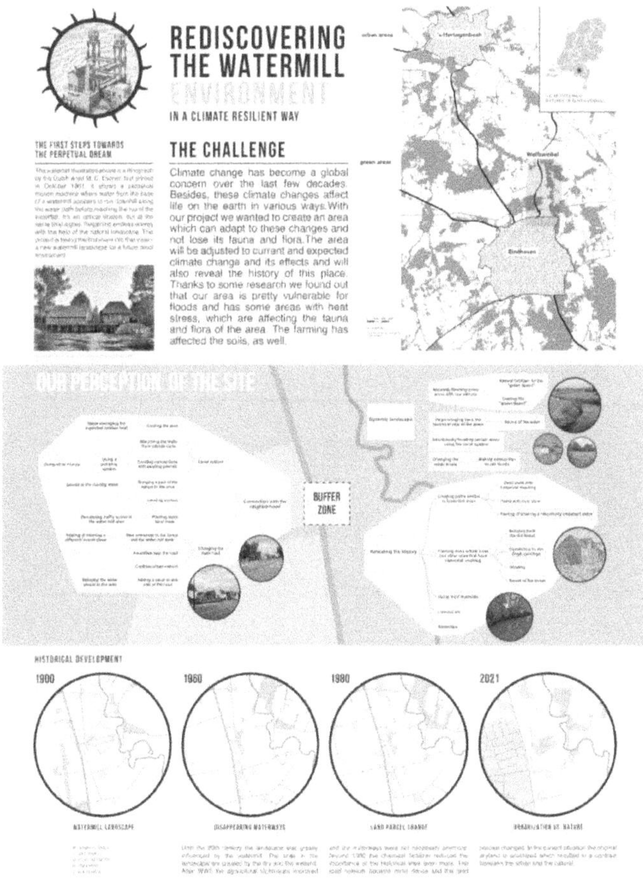

Figure 7.2a. *Historical analysis and site perception of Wolfswinkel watermill landscape (work by W. Heemskerk, M. Lamanna, M.B. de Klijn, D. Hasan, E. Pilatou, Z. Huang). For a color version of this figure, see www.iste.co.uk/curulli/heritage.zip*

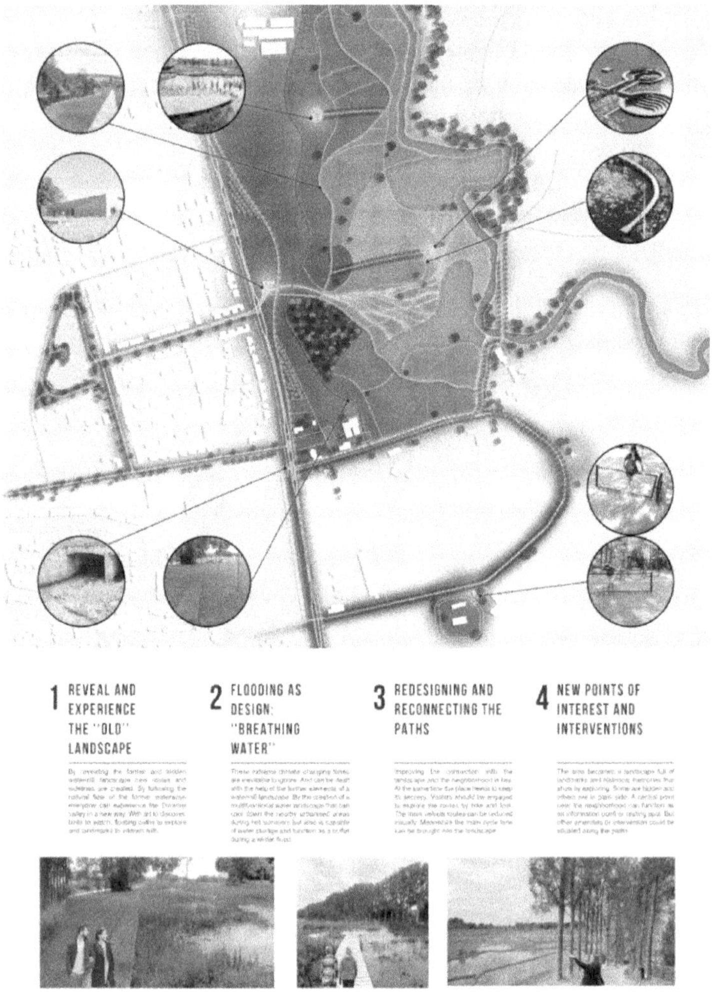

Figure 7.2b. *Design proposal to integrate the preservation of the historic landscape with water engineering for flood prevention (work by W. Heemskerk, M. Lamanna, M.B. de Klijn, D. Hasan, E. Pilatou, Z. Huang). For a color version of this figure, see www.iste.co.uk/curulli/heritage.zip*

Another proposal drew from landscape ecology and local cultural history to redevelop the watermill landscape. The project idea was to restore biodiversity and prevent flooding by re-establishing ecological connections, places of interaction, observing/learning from the landscape and retaining the excess of water by developing new floodplain meadows and wetlands (Figure 7.3).

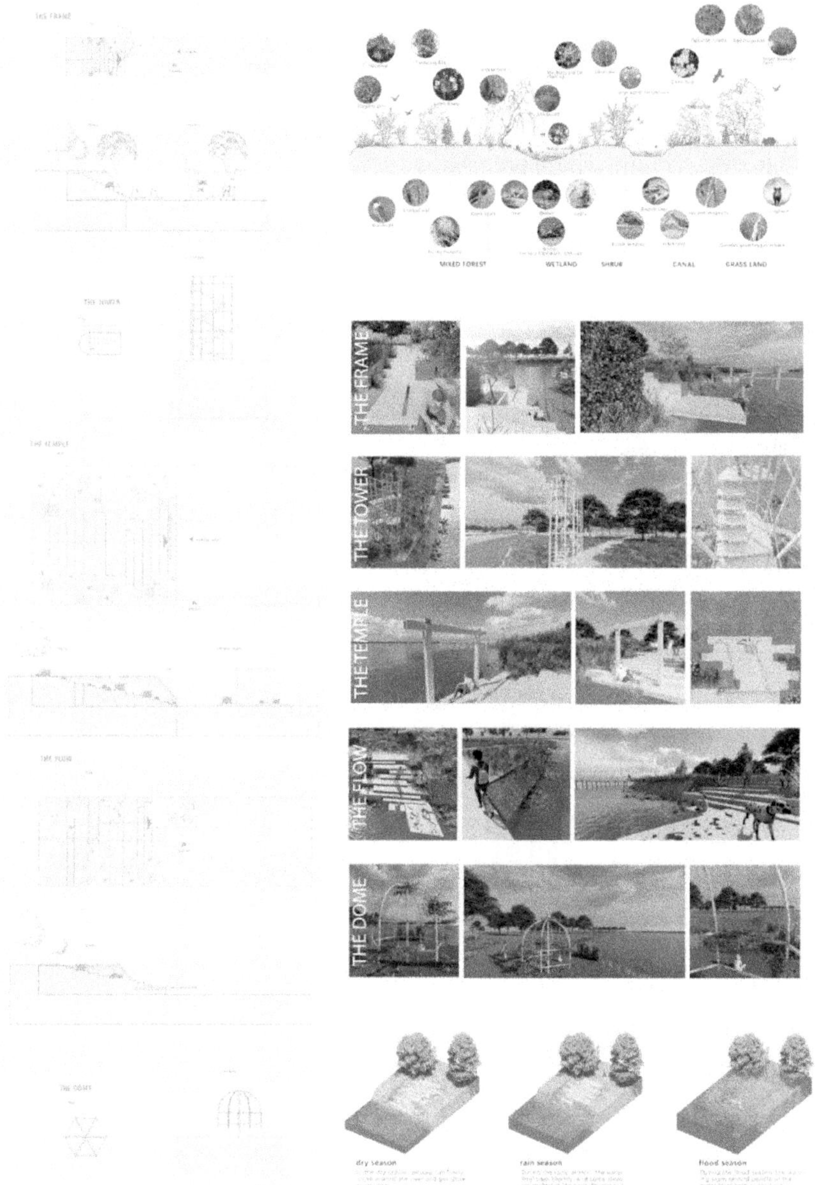

Figure 7.3. *Proposal for developing new floodplain meadows and wetlands (work by D. Çakir, C. Choveau, A. Di Stasio, Hugo-Haoran Hu, Siepan Khalil, Liu Xiong). For a color version of this figure, see www.iste.co.uk/curulli/heritage.zip*

In the "Vulnerability" workshop held in Molise in September 2022, site analysis was carried out by combining traditional database systems with GIS location data. By overlapping GIS hazards data (landslide, flood and fire) on land use maps, students detected potential places for interaction between hazards and heritage areas/objects and where to sketch alternative scenarios of prevention/reduction of heritage risks. For instance, a proposal integrated climate data forecasts with tourism strategies and focussed on people's experience of vulnerability. It developed a "risk awareness route" connecting minimal interventions that showed the extreme risks of the detected areas. Slow tourism developed on the route, which combined experiences, learning and pleasure, and it was meant as a tool to facilitate both behavioural change and societal support for climate actions (Figure 7.4).

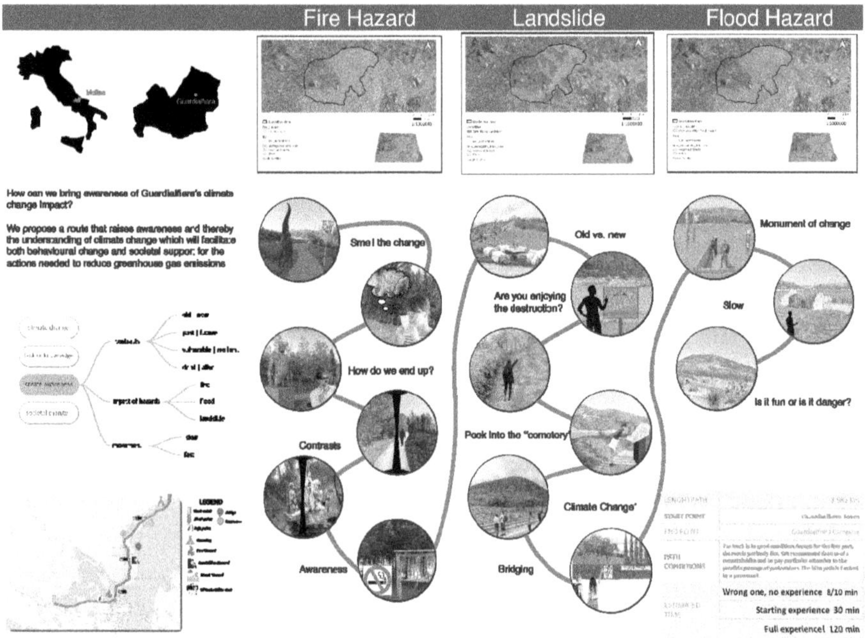

Figure 7.4. *Proposal to raise awareness of site's vulnerabilities through slow tourism (work by A. Simeonov Georgiev, D. Ricci, E. Sultan Kaya, E. Vergalito, M. Carrelli, M. Ziccardi, M. Kałuzny, R. Hutten). For a color version of this figure, see www.iste.co.uk/curulli/heritage.zip*

Another proposal drew from landscape and cultural geography disciplines and combined processes and functioning of nature with social experiences of transformation in the local communities. The work proposed site-specific nature-based solutions (NbS) to develop a resilient social-cultural heritage relationship that addressed the sense of belonging, preserved the identity of the place and provided optimized connectivity and mobility for the future (Figures 7.5(a) and (b)).

Figure 7.5a. *Assessment of climate change impacts on the cultural heritage of Agnone, Italy (work by S. Scoccola, E. Ceccacci, M.I. Spada, W. Kajda, S. Koržynska, H. Mahmoud, A. Mustafa, B. Öncül). For a color version of this figure, see www.iste.co.uk/curulli/heritage.zip*

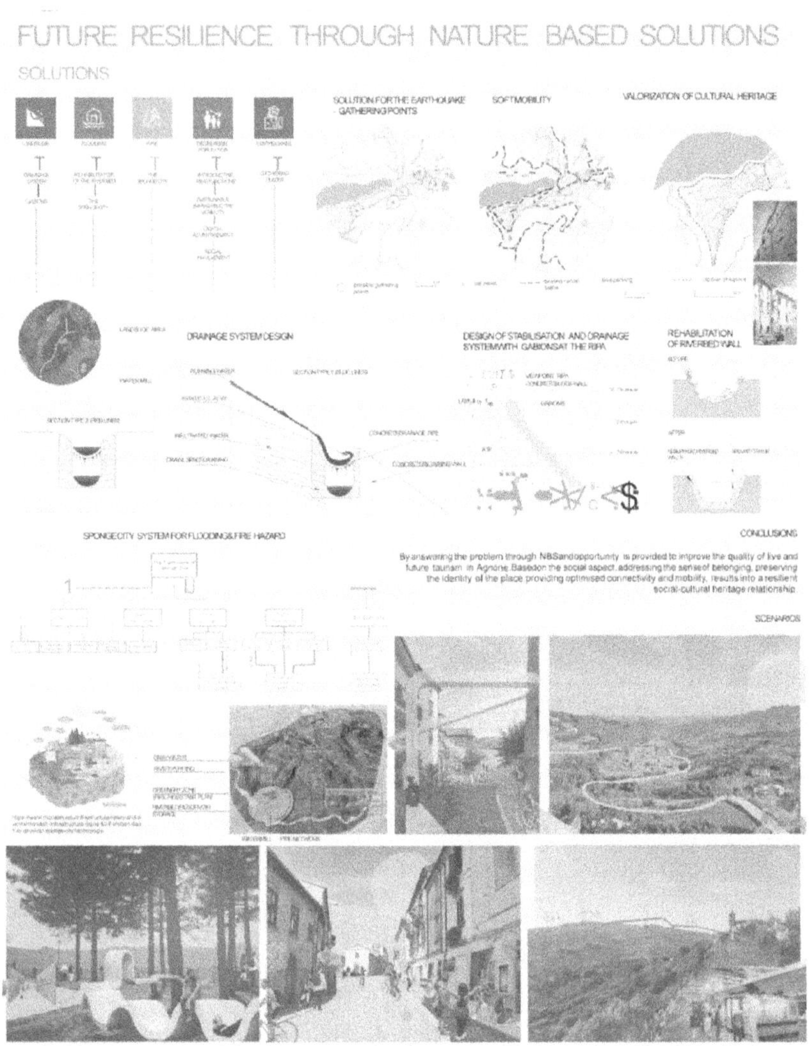

Figure 7.5b. *Proposal for nature-based solutions to climate change impacts in Agnone, Italy (work by S. Scoccola, E. Ceccacci, M.I. Spada, W. Kajda, S. Koržynska, H. Mahmoud, A. Mustafa, B. Öncül). For a color version of this figure, see www.iste.co.uk/curulli/heritage.zip*

Post-design phase

The design-led projects were coherently presented in verbal forms and visualized on boards. Results were discussed with visiting academics, professionals and representatives of local authorities/external partners. Proper presentations of design propositions were critically reviewed and appraised.

The criteria of the assessment focused on research quality, such as the documentation of the research process in visuals and findings, reflection on independent questioning by the students themselves, the design qualities (rigor, innovation, spatial dimensions, applicability in similar cases), the tools and techniques used, presentation and contextualization. The social and academic impact of the workshops and the new knowledge developed during the design process becomes available for a wider audience through reports and panels, following the route of academic publishing and debate.

7.5. Lesson learned: advancing heritage critical thinking through interdisciplinary teaching

Effective teaching on education for climate resilient heritage requires innovative learning methods and interdisciplinary approaches to develop students' skills toward comprehensive climate-sensitive and environmentally friendly professional actions. The processes and results of the two ISPs-design workshops (integrated with the assessment process of the e-CREHA online course) discussed in this chapter are successful examples of such innovative educational training and achievement of new competencies. The three overlapping phases of the design showed the interweaving of research and design and the interactive setting within which participants creatively exchange, collaborate and develop new knowledge together. Group discussions and presentations were used to monitor and adjust working practices during the workshops. The adoption of a multi- and interdisciplinary teaching allowed students to examine the link between heritage and climate change from multiple perspectives, leading to systematic efforts to synthesize and integrate the alternative perspectives into coherent analysis to forge viable responses/solutions.

Furthermore, interdisciplinary teaching helped students to go beyond the understanding of preconceived notions of heritage and opened up conversations to ideas from different disciplines, thus allowing all students to relate and contribute to the dialogue.

The design projects were not final spatial solutions, but conceptual innovations that showed a vast increase in learning and performance by the students. Throughout the workshop processes, students were able to build critical reflection skills, make links between heritage and relevant communities and stakeholders, and engage in ownership of the cultural heritage of their country and Europe.

Finally, the ISPs provided new lenses for a deep approach to heritage teaching for sustainable development. By combining diverse pedagogies with an interdisciplinary approach and focusing on problems from real cases, the ISPs have produced significant dissertation topics that help expand resilient thinking in heritage education. This is the final aim of the e-CREHA e-learning course.

7.6. Acknowledgments

We would like to thank the e-CREHA project partners for the development of the e-learning course, and for tutoring and organizing the project activities. A special thanks goes to all the students for their enthusiasm, dedication and discoveries that have made the success of the ISPs-design workshops.

The e-CREHA project was funded by the European Commission Erasmus+ KA203 program (no: 2020-1- NL01-KA203-064610).

7.7. References

de Aguilera, M. and Mendiz, A. (2003). Video games in education. *ACM Computers in Entertainment*, 1, 11.

Architects' Council of Europe (ACE/CAE) (2012). The architectural profession in Europe 2012 – A sector study commissioned by the Architects' Council of Europe. Report, Mirza & Nacey Research Ltd, Ford.

Asensio, M. and Young, C. (2002). The new visual literacy: The pedagogic value of streaming video. *European Distance Learning Network (EDEN) Conference*, Granada.

Bonk, C.J. and Graham, C.R. (2006). *The Handbook of Blended Learning: Global Perspectives, Local Designs*. John Wiley & Sons, Pfeiffer, San Francisco.

Brabec, E. and Chilton, E. (2015). Towards an ecology of cultural heritage. *Change Over Time*, 5(2), 266–285.

De Queiroz Barbosa, E.R., DeMeulder, B., Gerrits, Y. (2014). Design studio as a process of inquiry: The case of Studio Sao Paulo. *Rev. Lusófona Arquit. Educ./Archit. Educ. J.*, 11, 241–254.

European Council (2014). Strategic Agenda for the Union in Times of Change. Conclusions, 26/27 June. Report, European Council.

Gonzalez-Cacho, T. and Abbas, A. (2022). Impact of interactivity and active collaborative learning on students' critical thinking in higher education. *IEEE Revista Iberoamericana de Tecnologias del Aprendizaje*, 17(3), 254–261. doi: 10.1109/RITA.2022.3191286.

Graham, C.R. (2022). Exploring definitions, models, frameworks, and theory for blended learning research. In *Blended Learning: Research Perspectives*, Volume 3, Picciano, A.G., Dziuban, C.D., Graham, C.R., Moskal, P.D. (eds). Routledge, New York and Abingdon.

Holtorf, C. (2018). Embracing change: How cultural resilience is increased through cultural heritage. *World Archaeology*, 50(4), 639–650. doi: 10.1080/00438243.2018.1510340.

ICOMOS (2019). The future of our pasts: Engaging cultural heritage in climate action outline of climate change and cultural heritage [Online]. Available at: openarchive.icomos.org.

ICSM CHC (2022). Global research and action agenda on culture, heritage, and climate change [Online]. Available at: https://openarchive.icomos.org/id/eprint/2716/1/ICSM%20CHC-Global%20Research%20and%20Action%20Agenda%20on%20Culture%2C%20Heritage%20and%20Climate%20Change.pdf [Accessed 4 October 2022].

Mochizuki, Y. and Bruan, A. (2015). Climate change education in the context of education for sustainable development: Rationale and principles. *Journal of Education for Sustainable Development*, 9(1), 4–26. doi: 10.1177/0973408215569109.

Stupar, A., Mihajlov, V., Simic, I. (2017). Towards the conceptual changes in architectural education: Adjusting to climate change. *Sustainability*, 9(8), 1–19. doi: 10.3390/su9081355.

Šuligoj, V., Zavbi, R., Avsec, S. (2020). Interdisciplinary critical and design thinking. *International Journal of English Education*, 36(1), 84–95.

Tomey, A.M. (2013). Learning with cases. *The Journal of Continuing Education*, 34(1), 34–38.

UNESCO (2009). World Conference on Higher Education. The New Dynamics on Higher Education and Research for Societal Change and Development. UNESCO, Paris.

PART 2

Training on Heritage Vulnerability and Risk Assessment

8

Re-enabling Co-evolutionary Patterns to Reduce Landscape Vulnerability

Reframing planning approaches to the management of territorial resources could reduce the vulnerability of landscapes to the negative effects of climate change. In particular, planning could be oriented toward fostering recognition of territorial heritage by people living and acting in places in order to reactivate co-evolutionary dynamics of mutual adaptation between human beings and nonhuman environmental factors, that is, territorialization processes able to shape resilient landscapes.

8.1. Resilience and vulnerability

According to IPCC, resilience is:

> the capacity of interconnected social, economic and ecological systems to cope with a hazardous event, trend or disturbance, responding or reorganising in ways that maintain their essential function, identity and structure (IPCC 2022, p. 2920).

It is worth to note that in such a definition:

a) social, economic and ecological systems are considered interconnected (see section 2.2);

b) hazards are referred not only to events, but also to trends and disturbances.

Considering for now only the hazard component (b) of IPCC's definition of resilience, it is clear that in the scope of the climate-related science(s) the term

---

Chapter written by Luciano DE BONIS, Rossella NOCERA and Giovanni OTTAVIANO.

"hazard" does not refer only to the hazards of disasters but to a wider range of hazards, though it is also evident that it remains something related to risks; in particular climate risks, resulting

> [...] from dynamic interactions between climate-related hazards with the exposure and vulnerability of the affected human or ecological system to the hazards (IPCC 2022, p. 2921).

To understand resilience, it is therefore necessary to consider vulnerability, defined by IPCC as:

> the propensity or predisposition to be adversely affected. Vulnerability encompasses a variety of concepts and elements, including sensitivity or susceptibility to harm and lack of capacity to cope and adapt (IPCC 2022, p. 2927).

From the comparison of the previous definitions of resilience and vulnerability, it clearly emerges that while resilience could be defined as an ability to cope with something, vulnerability should be seen precisely as the lack of such ability, moreover in accordance with the common understanding of resilience and vulnerability as opposite terms.

On the other hand, IPCC recognizes that (i) vulnerability is a component of risk, but also, independently, an important focus; (ii) the concepts of adaptation, vulnerability, resilience and risk provide overlapping, alternative entry points for the climate change challenge (IPCC 2022, p. 43).

This means that we are allowed to consider the concept of vulnerability as not necessarily subordinated to an overarching concept of risk, for example, or to other "main" concepts such as resilience or adaptation (IPCC 2022, p. 43).

Furthermore,

> over the past several decades, approaches to analysing and assessing vulnerability have evolved: [from] an early emphasis on top-down, biophysical evaluation of vulnerability included – and often started with – exposure to climate hazards in assessing vulnerability [it has gradually emerged an] attention to bottom-up, social and contextual determinants of vulnerability [...] (Bergstrand et al. 2015; Rufat et al. 2015; Spielman et al. 2020; Taberna et al. 2020). Vulnerability is now widely understood to differ within communities and across societies, also changing through time (Kienberger et al. 2013; Jurgilevich et al. 2017) (IPCC 2022, p. 133).

8.2. Flexibility and "resiliencery vulnerability"

As mentioned above, the IPCC defines resilience as a capacity of interconnected social, economic and ecological systems. In this definition, a sort of dualistic conception remains, since the social and economic systems are evidently intended as "human", and the ecological one as "natural"; but at the same time the aspiration emerges to consider natural and human systems as interconnected. This aspiration is also evident from the recurring use of the term "socio-ecological system" (SES) in the 2022 Working Group II Contribution to the IPCC Sixth Assessment Report, where it is understood as:

> an integrated system that includes human societies and ecosystems, in which humans are part of nature. The functions of such a system arise from the interactions and interdependence of the social and ecological subsystems. The system's structure is characterised by reciprocal feedbacks, emphasising that humans must be seen as a part of, not apart from, nature (IPCC 2022, p. 2923).

Such a concept of SES shows a deeper relationship between natural and human systems than the generic "interconnectedness", although they still remain separate.

Separation instead completely disappears (though without any mystic indistinction) in the interpretation provided by Anderies et al. (2004), according to which a SES is "an ecological system intricately linked with and affected by one or more social systems", where "social" simply means "tending to form cooperative and interdependent relationships with others of one's kind", and social systems "can be thought of as interdependent systems of organisms". Consequently, the term "SES" can be precisely referred to:

> the subset of social systems in which some of the interdependent relationships among humans are mediated through interactions with biophysical and non-human biological units (Anderies et al. 2004).

On the basis of this very useful interpretation of SESs, we can now refer to the work of Gregory Bateson, to further move in the direction of considering the human as "a part of, not apart from, nature". According to Bateson (1972), the identification by Darwinian evolutionary theory of the unit of survival under natural selection with either the breeding individual or the family line or the subspecies, or some similar homogeneous set of conspecifics, contains "a very great error".

First of all because

> [...] the evolutionary unit is not homogeneous. A wild population of any species consists always of individuals whose genetic constitution varies widely. In other words, potentiality and readiness for change is already built into the survival unit (Bateson 1972, p. 458).

And above all because

> the flexible environment must also be included along with the flexible [not homogeneous] organism (Bateson 1972, p. 458).

Therefore,

> the unit of survival is a flexible organism-in-its-environment (Bateson 1972, p. 458).

Indeed, this unit – the living and evolving unit – can be seen as a cybernetic unit, in the sense that in order to interact with its ecosystem it is called upon to process all relevant information coming from the environment. And this information does not just travel along pathways internal to the borders of the epidermis. Unlike bio-energetic ecology, in informational or entropic ecology:

> the boundaries must enclose, not cut, the relevant pathways. Moreover, the very meaning of "survival" becomes different when we stop talking about the survival of something bounded by the skin and start to think of the survival of the system of ideas [information] in circuit (Bateson 1972, p. 467).

This means that the fundamental ecological units coincide neither with any individual organism of a certain species, nor with any conspecific aggregate of individuals, but rather with total cybernetic systems that process all the relevant information. That is to say with flexible organisms-in-their-environments that (a) require flexible delimitation according to their different behaviors; (b) do not therefore admit any sharp separation between organism and environment; and (c) are able to evolve precisely because they work via trial-and-error.

But what do we precisely mean when we speak about "flexibility", usually considered as the opposite of rigidity?

Again according to Bateson (1972), our civilization in general tends to immediately engulf all flexibility, that is, the "uncommitted potential for change", at the same time trying to limit this tendency by controlling the less rigid variables that invade the flexible space of the rigid ones, instead of favoring the necessary exercise of flexibility also of the latter.

In other words, our civilization tends to distribute flexibility in an inappropriate way, behaving like an acrobat who has his arms locked, but is "free" to fall, rather than like the real acrobat who, in order to keep himself dynamically stable on the rope, he moves his arms freely and continuously passing from one position of instability to another (Bateson 1972), that is, he simply uses all the necessary flexibility (not all the flexibility at his disposal) to keep "rigid" his condition of equilibrium, which Bateson very vividly presents as a continuous transition from one instability to another.

This means that, just as equilibrium should not be considered as the opposite of instability, flexibility should not be interpreted as the opposite of rigidity, but rather as a reciprocal complement to it, in the sense that it performs a collaborative function in maintaining the (necessary) condition of greater rigidity of some fundamental variables of the system.

But, what about the relationship between resilience and vulnerability, in turn usually considered as antonyms of each other?

Some scholars have already questioned the advisability of considering resilience as the positive part of vulnerability, and vice versa the latter as the "flip-side" of the former.

> In a certain way, it could be said that it is because one is vulnerable that one can be resilient: in the strictest meaning, there can only be resilience if there is an impact and a disruption, which, analytically, implies vulnerability. But in a larger sense, it is because a society or a territory is vulnerable that it will suffer crises and that it will have to face up, to adapt and to learn from the disaster [...] and become resilient in fact [...]. The passage from vulnerability to resilience cannot be limited to a simple semantic shift: the two terms are not interchangeable and the continuum question has to be explored as for what would make it possible to go from one to the other (cf. "resiliencery vulnerability" (Provitolo 2012)). [...] Moreover, it is not enough to transpose vulnerability's factors into resilience's field to understand hazard dynamics because a vulnerability factor can also be a resilience factor (Reghezza-Zitt et al. 2012, p. 13).

It can therefore be said that just as there is no equilibrium without instability, and there is no rigidity without flexibility, in the same way there is no resilience without vulnerability and vice versa.

8.3. Vulnerability of the Built Environment as natural/human system

As for the definitions of resilience and SES (see section 8.2), also in almost all IPCC definitions related to the Built Environment (BE), essentially two trends emerge: (1) the permanence of a traditional separation between the concept of Built Environment, intended as the man-made environment, as opposed to that of natural environment; (2) the simultaneous aspiration to overcome this separation, often expressed in a strong but still uncertain way.

By overcoming the traditional separation between natural and built environment, we do not mean here, by BE, the "anthropic" polarity opposed to the "natural" one of the global environment, but rather the scope of synthesis and symbiotic integration between the natural environment and the anthropic.

This is actually very close to the concept of landscape, and therefore allows us to extend the discourse from the BE to the BCE (Built Cultural Environment) as well as to consider BE and BCE as a whole (apart from the values we attribute to them).

In fact, if the "landscape" is to be conceived, as suggested by the European Landscape Convention (ELC, article 1, a.), "an area, as perceived by people, whose character is the result of action and interaction of natural and/or human factors", we must recognize that each element of the landscape – i.e. BE, i.e. (potentially) BCE – cannot be constituted, except in extreme cases, by natural factors alone or by human factors alone, but precisely by the result of their interaction (and intimate fusion).

Thus overcome, at least conceptually, the obsolete natural/human dichotomy; the real problem that remains can be expressed with the following question: how much of the present "built" natural/human environment reflects this idea of "a single system of environment combined with high human civilization" proposed by Bateson (1972)?

8.4. Vulnerability of the BCE as vulnerability of landscape heritage

According to the 1972 UNESCO Convention, the "World Cultural Heritage" includes monuments, groups of buildings and sites, which are of "outstanding universal value". Such a conception of cultural heritage has been severely criticized by Choay (1994), who derives it from the "cult of heritage" already reported by Riegl (1903), which manifests itself in the form of a kind of compulsive drive to conservative accumulation and to the narcissistic contemplation of the heritage, in which there is no relationship of interaction with the environment. Since it was precisely the interaction between human species and environment to produce in history what we now recognize as cultural heritage, Choay concludes that the compulsive drive toward conservative accumulation is a symptom of the progressive loss by our civilization of the ability to build and inhabit.

However, it must also be said that the concept of "continuing landscape", formulated more recently by UNESCO in the framework of the more general conception of "cultural landscape", seems able to overcome, at least in part, Choay's criticism. In fact,

> a continuing landscape is one which retains an active social role in contemporary society closely associated with the traditional way of life, and in which the evolutionary process is still in progress. At the same time it exhibits significant material evidence of its evolution over time (UNESCO 1992).

As such, continuing landscapes could be the most endowed with "resilient vulnerability" (Provitolo 2012) to climate change, in the sense that, as "living landscapes", they could be resilient precisely because they are somehow vulnerable (see section 8.2).

But, while very useful, the UNESCO concept of continuing landscape is severely constrained by being restricted to landscapes of "outstanding universal value". In fact, it is not only a question of protecting landscapes of cultural value (outstanding or not) from the negative effects of climate change, but also of the contribution that they can provide in terms of "resiliencery vulnerability", evidently the greater the more extensive the same landscapes are.

We believe therefore that the idea of continuing landscape should be extended, when the conditions are met, to the whole landscape as defined in the ELC, that is the entire territory (as perceived by people).

The strongest support for this possible generalization perhaps comes from the so-called Faro Convention, in particular its well-known definitions of "Cultural Heritage" and "Heritage Community" (Council of Europe 2005), together with the similar (but not identical) notion of "territorial heritage", provided by the Italian so-called "Territorialist School" and intended as:

> a heritage generated by long co-evolutionary processes between human beings and the environment, totally produced by human action and the result of an intimate integration of natural and cultural heritage (Magnaghi 1998).

By reformulating the concepts of cultural heritage and heritage community of the Faro Convention in the light of the "territorialist" notion of territorial heritage, the result could be the following (De Bonis 2020):

– The territorial heritage is the whole territory resulting from the continuous co-evolutionary interaction between man and the environment, in which widespread natural and cultural heritages are intimately integrated, that a group of people identifies as resources, regardless of who owns them, in relation to their own values, beliefs, knowledge and traditions.

– A community of territorial heritage is a group of people who attribute value to specific, multiple and integrated aspects of the territorial heritage, reproducing it through co-evolutionary interactions with the environment, within the framework of a collective action.

We believe that the conception of heritage embodied in the above definitions, especially the emphasis on its reproduction by heritage communities, may be able to counteract the progressive loss of competence in building and inhabiting mentioned by Choay, and can be the most suitable basis for addressing the vulnerability of heritage to climate change.

8.5. Climate resilient self-sustainable local developments

As stated above, we identify the BCE with continuing landscapes in a broad sense, including their archaeological (immovable) and architectural components, and above all not limiting us to consider those of (alleged) "outstanding universal value".

To this idea of the BCE, one of the concepts of the Contributions of Working Group II to the Sixth Assessment Report of the IPCC fits particularly well, namely the concept of "climate resilient development":

> climate resilient development (CRD) is a process of implementing greenhouse gas mitigation and adaptation options to support sustainable development for all [...]. Climate action and sustainable development are interdependent processes and climate resilient development is possible when this interdependence is leveraged [...]. In practice, adaptation, mitigation and sustainable development interventions are likely to be implemented in portfolio packages rather than as individual discrete options in isolation (IPCC 2022, pp. 2669, 2687).

Indeed, we believe that the most effective way to reduce vulnerability to climate change of the BCE (i.e. territorial heritage), or rather to increase its resiliencery vulnerability, consists of triggering climate-resilient development processes, which "flexibly" combine mitigation and adaptation to climate change with processes not of "simple" sustainable development, but rather of "self-sustainable local development", as defined by the Italian territorialist school (Tarozzi 1998), that is, with sustainable development processes achieved through the direct activation of local (heritage) communities that reestablish fruitful co-evolutionary relationships with their (territorial) heritage.

By interpreting these environmental relationships as a set of feedback processes, we can therefore state that they are, circularly, the cause and effect of the "continual process of choosing among and refining strategies of making a living (reproducing a way of life) in a changing world" (Knapp 2007) that characterize the history of human societies.

In this regard, it is worth noting that some settlement practices based on local knowledge and autonomous initiative have proven to be more effective in coping with changing environmental contexts than institutionalized interventions (White 1945).

It can therefore be argued that also in the context of climate change, which is a global phenomenon with effects on local environmental balances, it is necessary to implement "local mending" of the relational patterns that modernization has delocalized or globalized.

This also means rediscovering and continuously reconstructing the heritage of local knowledge that allows to develop (dynamic) practices of coevolution adaptive to (changing) contexts.

These local mending processes can benefit from the integration of (only) those nature-based solutions (NbS), which present a potential for the (re)production of BCE, and therefore of landscape. More specifically, interventions should be planned in which the NbS are not only a complement to the "gray" environment, but rather design solutions able to produce a greater integration between the act of living in a territory and the production of (continuing) landscapes, triggering an incremental process of dynamic (re)adaptation to the changing environmental context, for example, in the cases of the Xochimilco restoration project, in Mexico City, and the water retention landscape designed and realized in the village of Tamera, Portugal (Figure 8.1 and 8.2).

Xochimilco is one of the 16 boroughs of Mexico City, and it is best known as a swampy human settlement resulting from a historic process of anthropization of part of a vast wetland by the Aztec population. The main distinguishing feature of Xochimilco is the widespread network of water channels connecting the *chinampas*, the traditional Mesoamerican agriculture technique which reshaped the landscape of the Mexico Valley from the 15th century (Armillas Gil et al. 2016). The circumstances that took place through the centuries (from the Spanish colonization to the modernization and urbanization processes) deeply affected the SES of Xochimilco, producing water shortage and pollution, biodiversity threats, socioeconomic issues and eventually putting the local BCE at risk (Jiménez et al. 2020). Since the late Eighties of the 20th century, Xochimilco has been recognized as part of the UNESCO World Cultural Heritage (1987), as a Natural Protected Area by the Government (1992), as a site protected by the Ramsar Convention (2004), and as a Globally Important Agricultural Heritage System by the FAO (2017) (Pérez Galicia et al. 2019). All of these recognitions fostered the establishment of policies and actions aimed toward the protection of the site. Initially, they were mostly oriented to environmental conservation, and also set restrictions to the agricultural activities in the area. In more recent years, it became clear the role of traditional agriculture in shaping the specific SES that defines the natural and cultural (if detachable) local heritages. As a result, interventions were foreseen and implemented to promote and restore the chinampas' agricultural use (Zambrano et al. 2020). Among others, the most relevant actions consist in the reshaping of the borders of the channels between the chinampas as "green" treatment systems for the water to reduce its pollution; the restoration of the ancient streams to help feeding the channels network with clear water; the reconnection of the Rio Amecameca to the Xochimilco lake area to help regulating the water level (also through "gray" infrastructures); and the technologic enhancement of the systems for wastewater treatment.

Tamera is an ecovillage founded in 1995 in the Alentejo region of Portugal. Its focus is the development of a new planetary culture through a network of "Healing Biotopes", which carry out a nonviolent way of living. Starting from this core focus, the Tamera community (consisting of about 200 people) launched a set of actions aimed at achieving food, energy and water self-sufficiency, in the general framework of a renewed harmony between humans and nature. Beyond the ethical and spiritual reasons that motivated the community, and some criticalities about the global "sustainability" of Tamera's model (Esteves 2017; Mourato and Bussler 2019), what is relevant in the context of this chapter is how Tamera people acted in order to tackle the hazards related to climate change. Among a large number of measures that concur to the adaptation to CC, the most significant intervention has been the design and creation of a water retention landscape (WRL). Starting from

2005, the community reshaped the landscape to realize a system of lakes and ponds (19 in total), terraces and swales, and planted agroforests and orchards (Vizinho et al. 2021). All of these measures were aimed at strengthening the landscape capacity of retaining and managing the water from rainfalls to reduce the negative effects of extreme events, such as droughts, floods and soil erosion. The realization of the WRL is having positive contributions in terms of local microclimate, biodiversity, quality of life and food and water self-sufficiency (Esteves 2017; Vizinho et al. 2021).

Even if in very different contexts and backgrounds – the first is an ancient agriculture technique and settlement in a wetland, the second is a brand-new ecovillage in a dry rural territory – both cases are interventions aimed at reducing the vulnerability of the territory also through practices of interaction of human activities with the environment, and constitute responses to climate change effects that integrate NbS able to support processes of self-sustainable, climate resilient development.

Figure 8.1. *View of Xochimilco, Mexico. The set of interventions is related to the management of water resources to prevent floods, droughts, pollution, etc., also through integration between NbS and gray infrastructures. Image by LBM1948, via Wikimedia Commons, CC BY-SA 4.0*

Figure 8.2. *View of the water retention landscape of Tamera, Portugal. Image by Simon du Vinage, via Wikimedia Commons, CC BY 2.0*

8.6. Fundamentals of climate resilient territorial project

For territorial project educational purposes, the issues dealt with up until now can be summarized as indicated below.

First of all, the negative effects of climate variability do not consist only of catastrophic events but also of trends and perturbations, potentially able to affect almost the entire territory, thus reducing its habitability.

Moreover, the propensity to be adversely affected by events, trends and disturbances induced by climate variability, that is, the vulnerability to climate change, should no longer be seen as the contrary of the ability to cope with it, that is, as the opposite of resilience; for failure to identify the vulnerabilities of territorial elements to climate change also prevents the identification of the possibilities of resilient adaptation to its negative effects by the same vulnerable elements (and as such also resilient).

Nevertheless, to access a new and more effective conception of "resiliencery vulnerability" (Provitolo 2012), it is necessary to conceive the relationship between the human species and the environment in rigorously co-evolutionary terms, overcoming any residual dualism between them.

To achieve this goal, it is crucial not only to refer to the concept of SES, but also to interpret the latter as subsets of social systems in which the relationships between human beings are constitutively mediated by nonhuman abiotic and/or biotic elements (Anderies et al. 2004). Therefore, it is also necessary to reconceptualize the basic co-evolutionary units as nonhomogeneous (non-conspecific) units, always formed by different species, that cannot be delimited without including within them environmental factors that interact with human behavior, on which in turn the limits depend.

The similarity between this way of conceiving co-evolutionary units and the notion of landscape provided by the ELC is quite evident. But it is also possible to trace on these bases to a tendential coincidence not only the concepts of co-evolutionary units (CU) and landscape (L), but also those of BE and BCE, both between them and with the first pair of concepts (CU and L); provided that BE is no longer understood as the "anthropic" polarity opposed to the "natural" one, but precisely as the scope of synthesis and symbiotic integration between the natural and anthropic environments.

In other words, for the purposes of reducing vulnerability to climate change through territorial projects, it is legitimate to consider the BCE as a landscape, or better yet as a "continuing landscape" (UNESCO 1992) that is constantly evolving, which also includes archaeological (immovable) and architectural components. Even better, it is absolutely necessary to consider the architectural heritage and the immovable archaeological one as completely "immersed" in continuing landscapes, not necessarily of outstanding universal value (UNESCO 1992) but rather extended, barring exceptions, to the entire territory.

More precisely, starting from the excellent basis constituted by the notions of cultural heritage and cultural heritage communities provided by Faro Convention (Council of Europe 2005), it is necessary to move toward a conception of Territorial Heritage (Magnaghi 1998) and Territorial Heritage Communities, respectively, intended as the whole territory resulting from the continuous co-evolutionary interaction between man and the environment, and as groups of people who attribute value to specific, multiple and integrated aspects of the territorial heritage, reproducing it through co-evolutionary interactions with the environment, within the framework of a collective action (De Bonis 2020).

In fact, such a heritage conception is the most suitable to support these forms of climate resilient development (CRD) recommended by IPCC and intended as interventions that combine adaptation, mitigation and sustainable development (IPCC 2022). Moreover, it is also able to further evolve the concept of CRD toward the one proposed here of Climate Resilient Self-Sustainable Development (CRSSD), in which sustainable development processes are achieved through the direct activation of local (heritage) communities that reestablish fruitful co-evolutionary relationships with their own (territorial) heritage (see section 8.5).

There are already some promising examples of practices oriented toward forms of CRSSD, also capable of integrating those types of NbS that have the greatest potential for the (re)production of territorial heritage (see section 8.5). Those examples, together with the conceptual fundamentals summarized here, formed the learning framework for the experiments of territorial project conducted around the case study of the Molise region in Italy through extensive use of GIS technologies (see below).

8.7. A case study from the Molise region, Italy

In the learning framework described above, project works were developed by students involved in the e-CREHA workshop "Climate impacts and heritage vulnerabilities" (held in Termoli, Italy, September 13–22, 2022), as well as in the classes of "Territorial project" and "Geomatics" of the University of Molise's Master's degree in civil engineering, in form of CRSSD scenarios of some selected areas of Molise region.

The progressive definition of the intervention hypotheses started from an analysis of the regional territory aimed at selecting, through the use of various functions of geographic information systems (GIS), the areas where the spatial relationship between climate-related environmental hazards and immovable cultural heritage (archaeological, architectural and landscape) is more evident.

Students implemented a GIS that integrates the existing data about landslide, flood and fire hazards in the Molise region, as well as about the recognized heritage of four selected areas. All data were extracted from regional planning documents and institutional repositories, and in some cases further processed to make them usable for the project work. For instance, georeferencing and vectorizing of some raster data have been done, and selection by attributes of the specific categories of (immovable) heritage to be considered for the territorial project, among all vector data about protected heritage. Moreover, data about the land use from the Corine Land Cover inventory (III level) were integrated in the GIS to support the recognition of landscape patterns and their reinterpretation for the purposes of the project work.

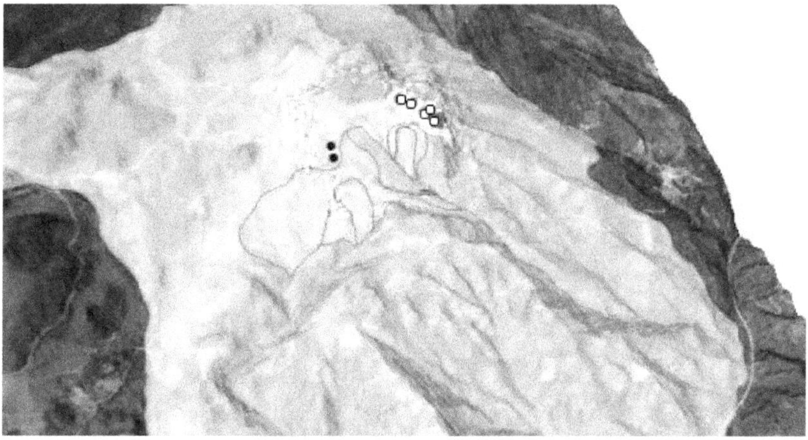

Figure 8.3. *Overlay of data about fire hazard (lighter part of the territory), landslides (outlined part of the lighter area), archeological heritage (black points), and architectural heritage (white points). Elaboration by G. Ottaviano of the original work by M. Carrelli, E. Ciccone, R. Cipriano, V. Cirocco, U. Corvino, R. Poce, M. Ranallo, S. Scoccola, M.I. Spada, G. Valiante, G. Vesce, F. Zappitelli and M. Ziccardi*

The overlaying of data concerning landscape, architectural and archeological assets, as well as concerning hazards and land use, made it possible to identify in broad terms the interactions between human activities and the immovable cultural heritage in the four selected areas. Students then tried to hypothesize, on the basis of the available studies, how climate change could affect this heritage, and consequently how to improve the resilience of landscape, understood as a whole as a BCE (see section 8.3 and 8.5). During the e-CREHA workshop, in particular, students developed intervention hypotheses to apply the general approach to the case study areas. Of all the proposals put forward, the ones that showed greater consistency with the CRSSD approach (see section 8.5) concerned the remodeling of part of the territory through the creation of terraces, and the use of appropriate NbS for water management, both aimed at reducing landscape vulnerability to landslides and floods, as well as enhancing agricultural production (Figure 8.5).

The same CRSSD approach was further developed during the aforementioned MSc classes in civil engineering. In particular, some improvements regarded a stronger consideration of the land use "as is" in the case study areas, and how it could be modified to allow a broader resilience of the territory to climate change effects. The more relevant actions in this direction regarded the change of crop typology in one of the study areas to reduce the vulnerability to landslides while increasing the economic revenue, and the reforestation of another area, to better manage the water from rains while triggering a local production of food and timber.

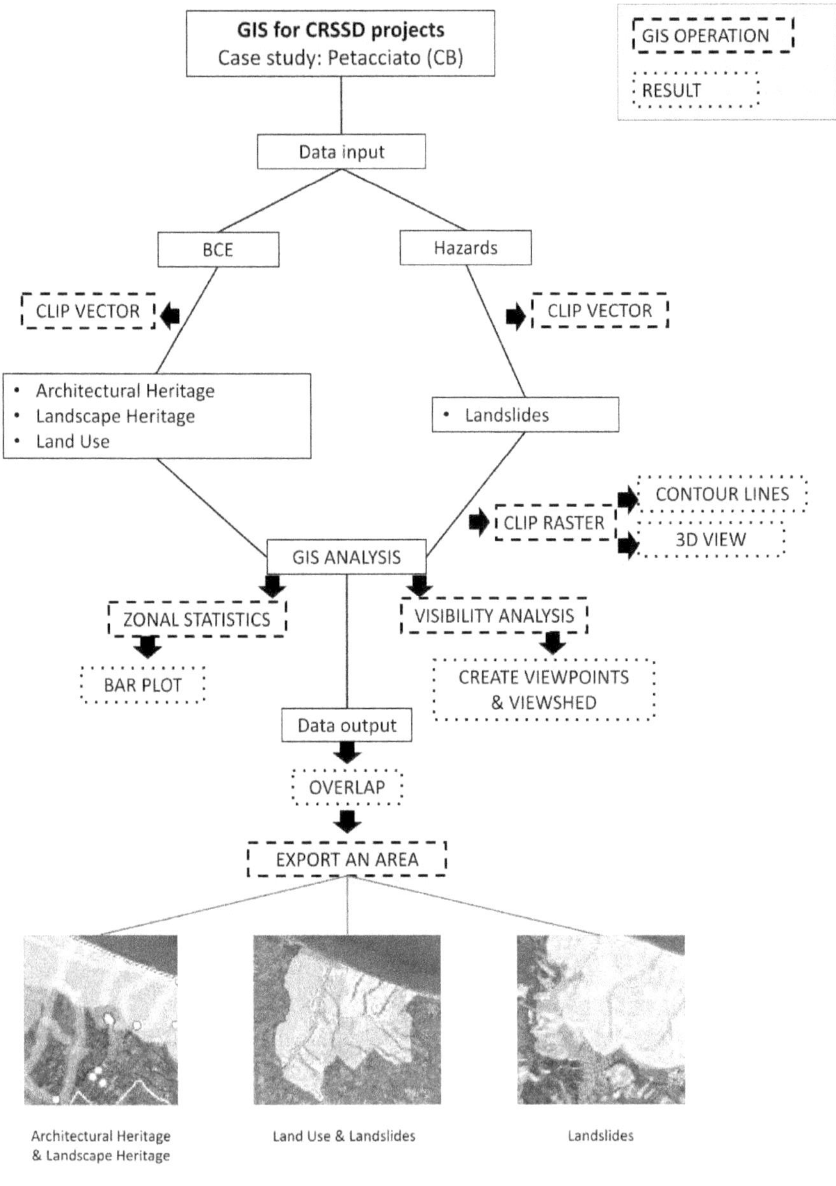

Figure 8.4. *Schematization of the GIS structure and operations made for the definition of the CRSSD project hypotheses in the framework of the e-CREHA workshop. Elaboration by G. Ottaviano of the original work by R. Cipriano and M.I. Spada*

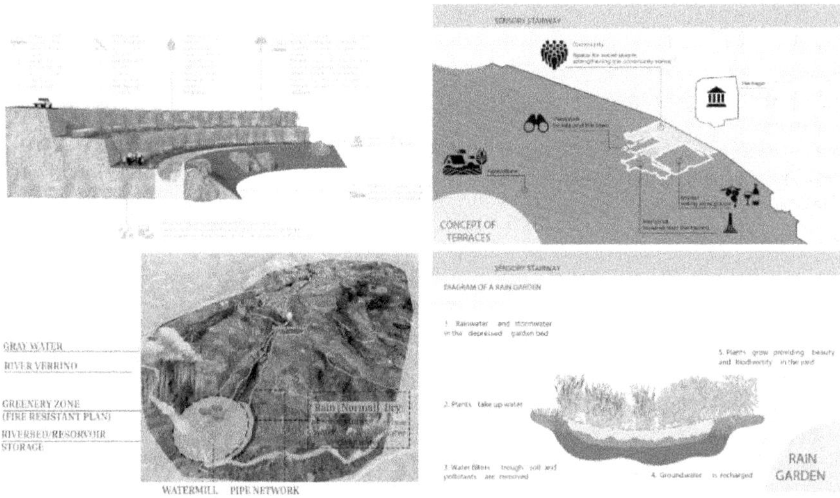

Figure 8.5. *Elaborations from the e-CREHA workshop "Climate impacts and heritage vulnerabilities". Tutors: S. Acar, S. Berahman, B.P. Bontchev, I. Curulli, L. De Bonis, E. Grodzka, P. Guzman, A. Leuzzo, R. Nocera, P. Gürol Öngören, G. Ottaviano, A. Özbek, M. Pelczarski, V. Terzieva and B. Yılmaz. Students: M. Buursma, M. Carrelli, E. Ceccacci, R. Cipriano, D. Dobrasiak, U. Doğanay, G.C. Dreliosi, A.S. Georgiev, R. Hutten, W. Kajda, M. Kałużny, E.S. Kaya, S. Kirsten, S. Korżyńska, T.P. Kostov, H. Mahmoud, A. Mustafa, S. Okur, B. Öncül, M. Ranallo, D. Ricci, S. Scoccola, M.I. Spada, T. Stoynova, S. Tanrıverdi, E. Vergalito, G. Vesce, F. Zappitelli, Y. Zhou and M. Ziccardi*

The application of the CRSSD approach in the described education context produced some interesting results in terms of definition of solutions consistent with the BCE of case study areas. On the other hand, it seemed quite challenging for the students to reach the above results, as the largest part of records and literature regarding measures aimed at reducing vulnerability to climate change effects does not integrate SSD practices. These findings clearly show the need to stress the CRSSD approach in project education in order to support the further elaboration and realization of interventions that contribute to reducing the negative effects of climate change on immovable cultural heritage while triggering retro-innovative ways of interacting with the territory able to generate and constantly regenerate continuing landscapes.

8.7. References

Anderies, J.M., Janssen, M.A., Ostrom, E. (2004). A framework to analyze the robustness of social-ecological systems from an institutional perspective. *Ecology and Society*, 9(1), 18.

Armillas Gil, I., González Pozo, A., Rodríguez Sánchez, L.C. (2016). Origins and evolution of the Chinampa landscape. In *The Chinampas of Xochimilco at the Start of the XXIst Century: An Initial Catalogue*, González Pozo, A. (ed.). Universidad Autónoma Metropolitana, Mexico City.

Bateson, G. (1972). *Steps to an Ecology of Mind*. Jason Aronson Inc., Northvale/London.

Bateson, G. (1979). *Mind and Nature: A Necessary Unity*. Bantam Books, New York.

Choay, F. (1995). Sept propositions sur le concept d'authenticité et son usage dans les pratiques du patrimoine historique. In *Conférence de Nara sur l'authenticité dans le cadre de la convention du patrimoine mondial : Nara, Japan, 1–6 November 1994*, Larsen, K.E. (ed.). Tapir Publishers, Trondheim.

Council of Europe (2005). *Council of Europe Convention on the Value of Cultural Heritage for Society*. Council of Europe Treaty Series, No. 199.

De Bonis, L. (2020). Processi di riterritorializzazione dell'area appenninica. In *Il governo del territorio montano nello spazio europeo. Innovare gli sguardi e gli strumenti per lo sviluppo sostenibile della montagna*, Cepollaro, G. and Zanon, B. (eds). Edizioni ETS, Pisa.

Esteves, A.M. (2017). "Commoning" at the borderland: Ecovillage development, socio-economic segregation and institutional mediation in southwestern Alentejo, Portugal. *Journal of Political Ecology*, 24(1), 968–991.

European Commission, Directorate-General for Research and Innovation (2015). Towards an EU research and innovation policy agenda for nature-based solutions & re-naturing cities: Final report of the Horizon 2020 expert group on "Nature-based solutions and re-naturing cities". Publications Office of the European Union, Luxembourg.

IPCC (2022). *Climate Change 2022: Impacts, Adaptation, and Vulnerability. Contribution of Working Group II to the Sixth Assessment Report of the Intergovernmental Panel on Climate Change*. Cambridge University Press.

Jiménez, M., Pérez-Belmont, P., Schewenius, M., Lerner, A.M., Mazari-Hiriart, M. (2020). Assessing the historical adaptive cycles of an urban social-ecological system and its potential future resilience: The case of Xochimilco, Mexico City. *Regional Environmental Change*, 20, 7.

Knapp, G. (2007). Human ecology. In *Encyclopedia of Environment and Society*, volume 3, Robbins, P. (ed.). SAGE Publications, Thousand Oaks.

Magnaghi, A. (1998). Il patrimonio territoriale: un codice genetico per lo sviluppo locale autosostenibile. In *Il territorio degli abitanti: società locali e autosostenibilità*, Magnaghi, A. (ed.). Dunod, Milan.

Mourato, J. and Bussler, A. (2019). Community-based initiatives and the politicization gap in socio-ecological transitions: Lessons from Portugal. *Environmental Innovation and Societal Transitions*, 33, 268–281.

Pérez Galicia, A., Pérez Campuzano, E., Delgado Campos, G.J. (2019). La configuración turística de las periferias urbano-rurales mediante estrategias gubernamentales. El caso de Xochimilco, Ciudad de México. *Revista Investigaciones Turísticas*, 17, 126–148.

Provitolo, D. (2012). The contribution of science and technology to meeting the challenge of risk and disaster reduction in developing countries: From concrete examples to the proposal of a conceptual model of "resiliencery vulnerability". In *Technologies and Innovations for Development*, Bolay, J.C., Schmid, M., Tejada, G., Hazboun, E. (eds). Springer, Paris.

Reghezza-Zitt, M., Rufat, S., Djament-Tran, G., Le Blanc, A., Lhomme, S. (2012). What resilience is not: Uses and abuses. *Cybergeo: European Journal of Geography, Environment, Nature, Paysage*, 621 [Online]. Available at: http://journals.openedition.org/cybergeo/25554 [Accessed August 1, 2023].

Riegl, A. (1903). *Der moderne Denkmalkultus, sein Wesen und seine Entstehung*. W. Braumüller, Vienna/Leipzig.

Taleb, N.N. (2012). *Antifragile: Things that Gain from Disorder*. Random House, New York.

Tarozzi, A. (1998). Autosostenibilità: una parola chiave e i suoi antefatti. In *Il territorio degli abitanti. Società locali e autosostenibilità*, Magnaghi, A. (ed.). Dunod-Masson, Milan.

UNESCO (1992). Operational guidelines by the world heritage committee at its 16th session. Document WHC-92/CONF.002/12, UNESCO, Santa Fe.

Vizinho, A., Cabral, M.I., Nogueira, C., Pires, I., Bilotta, P. (2021). Rural renaissance, multifunctional landscapes, and climate adaptation: Trilogy proposal from grassroots innovation and participatory action research projects. In *Handbook of Climate Change Management*, Leal Filho, W., Luetz, J., Ayal, D. (eds). Springer, Cham.

White, G.F. (1945). Human adjustment to floods. A geographical approach to the flood problem in the United States. Research paper no. 29, University of Chicago.

Zambrano, L., Rivas, M.I., Uriel-Sumano, C., Rojas-Villaseñor, R., Rubio, M., Mena, H., Vázquez-Mendoza, D.L., Tovar-Garza, A. (2020). Adapting wetland restoration practices in urban areas: Perspectives from Xochimilco in Mexico City. *Ecological Restoration*, 38(2), 114–123.

9

Lebanon: Landscape at Risk

9.1. Introduction

This chapter addresses the challenges and related risks that Lebanese landscapes are subjected to. A short introduction to the position of the landscape discourse in the country will be followed by a description of the critical situation in which Lebanon finds itself. The country faces multiple unprecedented crises, including socio-political, economic and infrastructural challenges amid a pandemic. The situation has been further exacerbated by the Beirut Port explosion on August 4, 2020, which shook the city's foundations, injured and killed many, damaged the port and properties at the heart of the city and rendered thousands homeless. The risks reported in the paper are related to the experiences conducted at the American University of Beirut with the students in the Landscape Design and Environmental Management department (LDEM). During the 2019–2020 academic year, professors and students who were a part of the Landscape Capstone Project, I and II (LDEM 241-242) researched adaptation and mitigation measures to ensure that Lebanese landscapes and societies better respond to the inevitable threats and consequences of weather-related disasters. In developing countries, more than in other parts of the world, landscapes are at high risk of socioenvironmental decay. The pressure of urbanization, energy demand, pollution and the waste crisis, amplified by the high influx of displaced and refugee communities, are compromising Lebanese land. To respond to some of the pressing 2030 Sustainable Development Goals (SDG), the LDEM instructors, Maria Gabriella Trovato, Mona Khechen and Balsam El Ariss selected the theme "Landscapes at Risk" as a topic for the LDEM 241-242. This chapter will use the course experience and the description of students' project results to dig into the condition of risky Lebanese landscapes and present some future scenarios challenging climate change and cultural landscapes.

Chapter written by Maria Gabriella TROVATO.

9.2. The Landscape

"Landscape" is a ubiquitous word in English and other European languages with origins in Anglo-German, dating back to circa 500 CE in Europe (Taylor 2022). However, it means different things to different people, and it is frequently treated as interchangeable with various other words, notably including nature, countryside, scenery, place and environment (Fairclough et al. 2018). In 1979, Meinig stated that landscape is related to, but not identical to nature; every landscape is a scene, but is not identical to the scenery; landscape is all around us, but is not identical to the environment; landscapes are related to, but not identical to places; landscape is a portion of the earth's surface, related to, but not identical to the region, area or geography.

Before the European Landscape Convention (ELC), several theories were suggested for the contributions of different aspects of nature to other aspects of physical (Ulrich 1997), social (Alexander et al. 1977; Lennard 1987; Lynch 1960, 1981) or psychological well-being (Ulrich 1983, 1993; Kaplan and Kaplan 1989; Ulrich et al. 1991; Wilson 1993; Kaplan 1995). In 2000, the ELC came into force, and allowed the landscape to be seen as an important public interest and a central component of people's well-being and quality of life (Selman 2004; Jones and Stenseke 2011). Moreover, the European Landscape Convention emphasizes the active engagement of people who, when recognizing an area as a landscape, also constitute it as such and assign landscape values to it (Jørgensen et al. 2016). In Arabic, the term landscape does not have a direct equivalent. Nevertheless, according to Makhzoumi (2002), several terms have been traditionally used by Middle Eastern people, such as hada'ik (gardens), fadha'at kharijia (outdoor spaces) and bi'a (environment), connoting one or more meanings of landscape. The shortcoming is partly due to the complexity of the English word, which makes it impossible to combine the different meanings into a single translation, but is mainly due to the difficulties inherent in transferring a complex concept, such as "landscape", from one sociocultural context to another (Makhzoumi 2002). Landscape has then been variously translated as "natural scenery" (Makhzoumi 2002), "land scenery" (Nasr and Khatib 1985; Baalbaki and Baalbaki 1997) and a "view of the countryside" (Makhzoumi 2002). Therefore, in the Middle East, "landscape" can be defined in two different ways: as a physical entity, a piece of the Earth's surface and its system of living, non-living and human components, and as a social and cultural construction, signifying how people engage with their world at a specific time and place (Makhzoumi 2002).

In this part of the world, where the claim over natural resources has fueled political and economic instability and unfinished wars on control of rich territories, the landscape discourse has long been relegated to urban beautification (Makhzoumi 2016). However, the region's fast development poses an urban footprint pressure on the regional ecologies. The related urban livability challenges, overstressed by climate-related issues, have relegated the landscape and ecological debate to a secondary and unimportant level. Additionally, planning and decision-making are under government institutions. This colonial legacy continued to dominate postcolonial decades and resulted in weak local authorities in decision-making and responding to local needs. As a result, the public realm is increasingly suffering, and the needs of the locals and citizens are rarely heard (Makhzoumi and Al Sabagh 2018). Valuation of cultural landscapes is equally lost between environmental scientists that focus on measures for preventing soil erosion and mitigating pollution of water resources, and architects that prioritize the historical, vernacular built heritage in rural towns and villages (Makhzoumi et al. 2012). Indeed, echoing Makhzoumi's research, the landscape approach is especially relevant in the Middle East because it has the potential to sidestep linguistic shortcomings (Makhzoumi 2002), while eliciting a rich and layered understanding of landscape.

In this chapter, heritage and landscape are strictly related notions since the latter is considered a process in which humans create the cultural landscape. Landscape is a place of practice, sociocultural values and repositories of traditions. Moreover, we agreed with Roe and Taylor (2014) in considering our heritage and cultural landscapes as living entities, constantly changing to respond to needs and contemporary challenges through an interactive process between natural systems, including human interventions and actions. Therefore, the Lebanese landscape is approached and considered as the composition of material and tangible elements with intangible components embedded in the history of the places and in the memory of the people who created them. It is interpreted as the outcome of people – environment co-evolution (Makhzoumi 2009); an entity we live in and pass through every day, affecting our lived experiences tangibly and intangibly (Roe and Taylor 2014) as everyone's heritage and an essential part of our patrimony (Lowenthal 2007). Unfortunately, like most of the Eastern Mediterranean region, the landscapes in Lebanon are currently under-studied and called out for further examination. The recurrent and severe climatic and ecological threats coupled with the socio and political crisis have caused ceaseless stress on the land and its people (Snow and Kluiving 2022). Those threats required a different approach embedded in the belief of a continuous change in the relationship between humans, natural resources and events.

9.3. Lebanon

With a surface of 10.400 km^2, Lebanon is the second smallest country in the Middle East and the Arab World (Council for Development & Reconstruction 2005). Stretching for 225 km from north to south along the Mediterranean Sea, it is characterized by four main parallel types of landforms (Figure 9.1): a flat, narrow coastal strip along the Mediterranean Sea, the Lebanon Mountain up to 1000 m above sea level, a large high plateau, the Bekaa Valley, at around 900 m above the sea level, and the anti-Lebanon chain, which rises to 2800 m, stretching across the eastern border with Syria (FAO 2008).

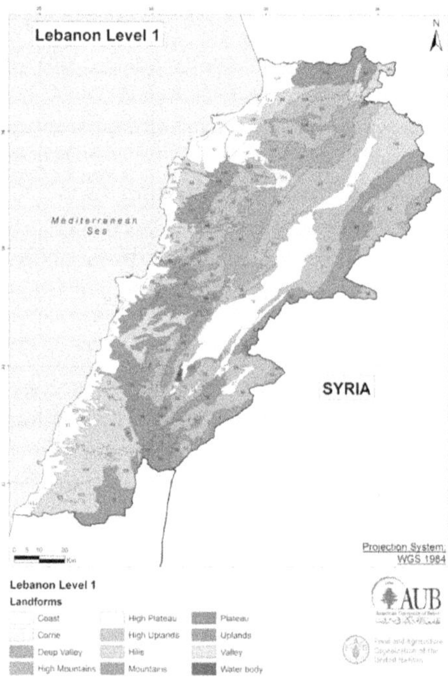

Figure 9.1. *Lebanon landforms. For a color version of this figure, see www.iste.co.uk/curulli/heritage.zip*

Unique among the countries of the Middle East, Lebanon is a mélange of diverse peoples, cultures, and religious creeds. For centuries, it lay at the crossroads of civilizations with a history marked by the ancient Egyptians, Phoenicians, Greeks, Romans, Byzantines, Arabs, European Crusaders, Mamluks and Ottomans (Wiener 2017). As a result, Lebanon's cultural landscape is the hallmark of the land of history and culture, an expression of a dynamic collaboration among natural and social identities. Unfortunately, Lebanon is experiencing an unprecedented economic crisis that has caused inflation to rise and essential public services to be cut dramatically (Anonymous 2022). This crisis is the product of its corrupted political system, which cannot deliver effective governance transparency and accountability. The 2019 protest movement revealed the deteriorated economic and sociopolitical dysfunction due to corruption and nepotism (Anonymous 2019). The emergence of the Covid-19 pandemic and the Beirut Port explosion in 2020 have exacerbated the already deteriorating economic situation, causing waves of extreme political unrest and government resignation. The blast, which damaged the country's trading hub and surrounding neighborhoods, caused shortages of basic food supplies, increased reliance on external or international aid, and a spike in Covid-19 cases, as many were left homeless. Hospitals were overwhelmed with casualties and sustained damage (Trovato et al. 2021).

Disasters have been central in the making of Lebanon since its inception. It consists of events that are (re)produced, maintained and magnified by contemporary neoliberal policies (Riachi 2022). Ecological disasters overlap with man-made and climate-related events. Recently, climate change interaction with the existing economic, political and environmental stressors is accelerating existing risks and creating new obstacles.

In the past few years, Lebanon has been heavily affected by climate change. We have witnessed a difference in the weather patterns that has reduced the total volume of available water snow coverage and increased drought periods. As extensively reported by the Lebanese Ministry of Agriculture (2003), the abrupt decrease of water in the last few decades has directly affected water availability for users, especially in the agriculture sector. It has been estimated (MoE 2016) that by 2040, temperatures will rise by 1°C on the coast and 2°C inland. The precipitation is expected to decrease by 10–20%, and less humid and sustainable warmer conditions will affect the warm and dry climates. Coastal areas will experience a rise in sea levels and seawater intrusion, exacerbating the ongoing salinization of coastal aquifers (Ward and Ruckstuhl 2017). Moreover, climate change will lead to a substantial decrease in snowfall (MoE 2011) and the altitude of ice. Those variations will affect the recharges of the rivers and groundwater. Rivers will endure a decrease in flows and extreme changes in the riverain socioenvironmental landscape. As a direct response to all these major threats, no significant legislation directly addresses climate change action in Lebanon, nor is climate change a mainstream legislative requirement (MoE/UNDP/GEF 2019).

At the Landscape Architecture and Ecosystem Management Department (LDEM) of the American University of Beirut, professors and students have been researching adaptation and mitigation measures that could ensure that Lebanese landscapes and societies better respond to the inevitable threats and consequences of weather-related disasters. Landscapes are at high risk of socioenvironmental decay. The pressure of urbanization, energy demand, pollution and the waste crisis, amplified by the high influx of displaced and refugee communities, are depleting natural resources, impoverishing the land and wrecking the innate human, non-human process. As a result, heritage is at risk of extinction. The studio course adopted the landscape approach with the aim of providing concepts and tools for managing heritage toward sustainable practices (Guzman et al. 2020).

This chapter reports on the experience conducted in 2019–2020 at LDEM as part of the Landscape Capstone Project I and II (LDEM 241-242). The studio aims to develop student independence in defining project issues and successfully integrate all aspects of the design/research proposal in a comprehensively developed project. To respond to some of the pressing 2030 Sustainable Development Goals (SDG), the instructors, Maria Gabriella Trovato, Mona Khechen and Balsam El Ariss, selected "Landscapes at Risk" as the course theme. Hence, the students focused on threatened landscapes, mainly landscapes of multi-performative natures. Students were tasked to select locations in Lebanon that could respond to the topic of the course. Approaches to the design, planning and management of the chosen sites used landscape as a process to protect the cultural and ecological infrastructures for an expanding human settlement. Landscape strategies have been explored to couple with the challenges of the sites proposing design interventions that suggest beneficial relationships between settlement form, culture, society and ecological environment. As a result, the student's work was a collection of local place-specific scenarios. In this chapter, three projects have been selected and described. Two students selected the water-related landscape as a case study and studied the risks of climate change on the cultural riverscape. The decrease in annual precipitation and related drought have altered the traditional relationship local communities used to have with the riverscape. Thoroughly analyzing the site's components, the students individuated the major stressor factors on the landscape as a cultural construct. They proposed viable scenarios for the future of the chosen areas. One intervention was located in a mountain region of Lebanon that has been interested in extensive mining activity since the last century. Moreover, the site is inside and at the border of the biosphere reserve in the Shouf.

9.4. Heritage, climate change and the landscapes at risk

The Cyprus Institute has reported that the Eastern Mediterranean and the Middle East (EMME) have now become a global climate change "hot spot". The gradual warming of temperatures and the decline in precipitation increase heat stress and reduce rainfall. As a result, there will be a significant reduction in freshwater availability for the EMME with significant socioeconomic repercussions (Lelieveld et al. 2012). Furthermore, projections suggest that the environmental variables associated with climate change in the region are putting the natural system at high risk, contributing to the degradation and loss of cultural landscapes, and threatening traditional cultural practices (Johnson and Vida 2020). Heritage and climate change have gained significant attention in recent studies conducted by academics, policymakers and managers (Harvey and Perry 2015).

In 2005, the World Heritage Committee widely discussed the impact of climate change on heritage sites, which led to the report and policy in 2007 and a volume on case studies in 2009 (Colette 2007, 2009). Successively, the interest in the topic increased within academic circles (e.g. Cassar 2005; Lemieux et al. 2011; Perry 2011, 2015; Barthel Bouchier 2013). In parallel, a growing body of research has reported on climate change's significant effects on the natural environment and human communities (Harvey and Perry 2015). There is a consensus that global climate variations and related risks affect communities' relationships with the environment. The earth is warming at a sustained rate. Global warming occurs worldwide, affecting the Mediterranean region more than other parts (Lionello and Scarascia 2018), with marked temperatures and precipitation changes. As a result, the magnitude and frequencies of extreme events are related to climate change increase, and unprecedented disasters threaten communities. Several areas in different countries have already been affected by catastrophes that significantly damaged natural and cultural heritage, expanding costs to societies and economies (Cacciotti et al. 2021). Indeed, we can enumerate some examples of the impact of extreme climate-related events on cultural landscapes, such as the sea-level rise in Venice (Ivajnšič et al. 2018); flood and coastal erosion in Mediterranean UNESCO World Heritage Sites (Reimann et al. 2018); changes in air and sea temperatures in Shark Bay in Australia (Strydom et al. 2020); variations in humidity; severe weather events such as hurricanes, storms and weathering on Puerto Rico's southeast coast (Keellings and Hernández Ayala 2019) and changes in soil and sediment conditions in Cyprus's Paphos area (Cuca and Agapiou 2018).

9.5. Method

The LDEM 241-242 undergraduate courses on Landscape Architecture followed a studio format. They adopted a methodology that entailed landscape analysis and characterization, planning and design experiments to produce strategies for future intervention in the context of landscapes, considering heritage and sociocultural expression. Field research and literature review of the design case study helped students to comprehend the site's history, physical, ecological and social processes, and constraints. The studio investigated the role of landscape architecture in dealing with contemporary global challenges and related risks on the communities and territories. To that end, a case study approach has been used, and the students have selected a series of locations in Lebanon to respond to the course topic. Certain questions trigged the landscape process and helped students in individuating challenges and framing their design narratives. Which are the elements of change? Why is the landscape at risk?

In this chapter, I present the results of three projects. Two focus on rivers and tackle social environmental justice issues and landscape recovery of cultural waterscapes. Students were tasked with assessing conflict interests on water sharing, quality and ecological resilience with flood control, public/private rights and access, pollution and public safety. The sites are in south Lebanon: Sayniq River in Saida, on the southern Lebanese coast, and Hasbani River, a major tributary of the Jordan River. The third project, "Ain Dara's Hidden Ecological Potential: The Quarry Park", is on recovering an abandoned quarry located at the border of a natural reserve.

9.5.1. *Case study*

9.5.1.1. *Sayniq River*

Watercourses in the Southern Lebanese coast have historically influenced the east-west alignment of agricultural lands and movement trunks linking the sea and the regions east of the mountains. In Saida, one of the major Lebanese cities, six watercourses intersect its territory with the Sayniq River in the southern part. The river flows from the spring Em Chammas in the Jezzine District and outflows into the Mediterranean Sea in the Saida District (Figure 9.2). It encompasses an industrial area, leisure spots like resorts and restaurants, agricultural zones and lands with cultural elements like mills and churches. The river was known for its swimming and gathering places and was also used to irrigate adjacent lands on both sides. This watercourse has almost dried out in the past few decades due to climate change and the urban development pressure uphill, becoming the repository of infrastructural and industrial sewages.

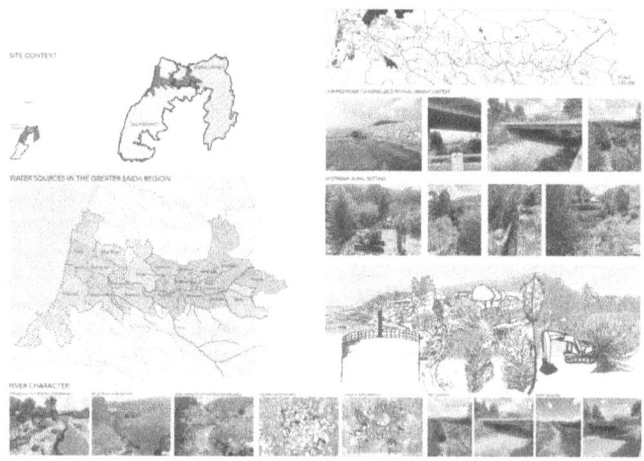

Figure 9.2. *The Sayniq River in Saida, South Lebanon (credit: Nour El Zein). For a color version of this figure, see www.iste.co.uk/curulli/heritage.zip*

Moreover, the depletion of the natural flow, the poor water quality and the new river land use determined a loss of the community's connection to the riverscape.

9.5.1.2. *Hasbani River*

In Southern Lebanon, the Hasbani River, a major tributary of the Jordan River, runs for 25 miles before reaching occupied Palestine. It is a regional watercourse that starts from Hasbaya village, crosses the borders into the Lake of Tiberias in Occupied Palestine and then feeds into the Jordan River (Figure 9.3). Passing the borders of two conflicting countries, the riverscape witnessed disputes or wars over water in the middle east. According to the Johnston Plan in the 1950s, Lebanon's rightful share of this river is 35 million m^3 in water volume. Instead, the country currently uses only 17 million m^3. Moreover, regional conflicts and poor management have led to the under-utilization of water resources and increased pollution. The topographical character of the waterscape and the lack of regulated conflicts between public and private interests have resulted in difficulties in accessing the river. This has created differentiated use and management of its shores, with a concentration on leisure activities in well-defined sections of the watercourse.

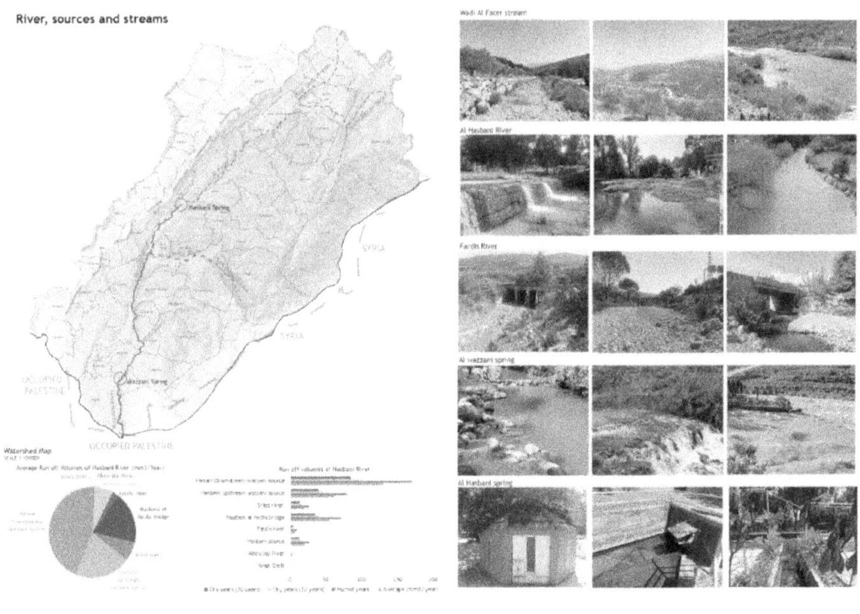

Figure 9.3. *The Hasbani River in South Lebanon (credit: Lama Amin). For a color version of this figure, see www.iste.co.uk/curulli/heritage.zip*

9.5.1.3. *Ain Dara quarry*

In Lebanon, quarrying activities wield pressure on soil and water resources, thus accelerating erosion processes and subsequent destruction of existing arable lands (Darwish et al. 2004). Moreover, mining activities alter pre-existing ecosystems, change landscape patterns and integrity (Jomaa et al. 2008), destroy natural habitats (Khater 2004) and increase dust emission, noise pollution and traffic to and from the quarry site. Ain Dara is a tiny village in the Shouf District of Mount Lebanon, surrounded by a pine forest. The quarries and crushing plants surrounding the village have tremendously damaged its environment, air quality and scenic beauty (Figure 9.4). Moreover, the Ain Dara quarry is located within the Shouf Biosphere Reserve's buffer zone. The activities have been carried out illegally since the end of the Lebanese civil war and were stopped in 2009. However, these practices have transformed the village's overall character and identity and threatened Ain Dara citizens' health.

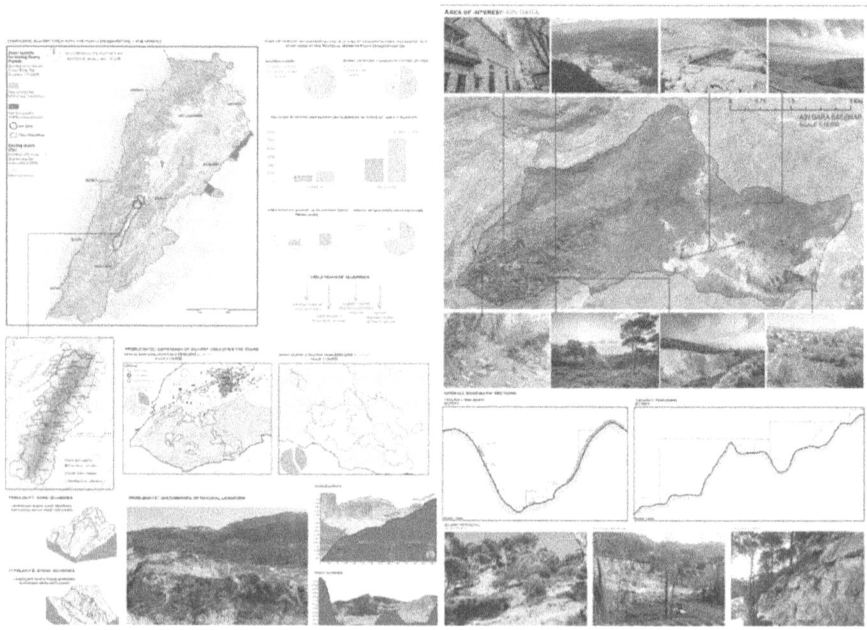

Figure 9.4. *The quarries in the Shouf district of Mount Lebanon. The Ain Dara quarries (credit: Julia Abi Saab). For a color version of this figure, see www.iste.co.uk/curulli/heritage.zip*

9.6. Discussion

Landscape architecture offers a proactive set of methodologies and tools for a sustainable and responsible regeneration of areas that are under pressure, at risk and in a condition of decay. Indeed, students' design proposals targeted sites with important and evident risks of loss of sociocultural and environmental components. Their proposed scenarios aim to recover the memory of past landscape practices and increase biodiversity. All over the world, mining activities deplete the land, impoverish the soil and cause serious health problems. Quarries can cause alterations, decay or loss of material and immaterial assets that characterize landscapes. They abruptly interrupt the continuity of habitats and cause soil erosion, air pollution and deterioration in water quality, but they also take a toll on people's overall health. The "Ain Dara's Hidden Ecological Potential: The Quarry Park" landscape project aimed to restore a disturbed ecosystem by addressing its environmental constraints and rehabilitating its ecological performance. It intended to emphasize the necessity of an intervention on a landscape at risk in the form of water management and vegetation techniques (Figure 9.5). Those techniques aimed to trigger a rehabilitation process within a landscape ecological context.

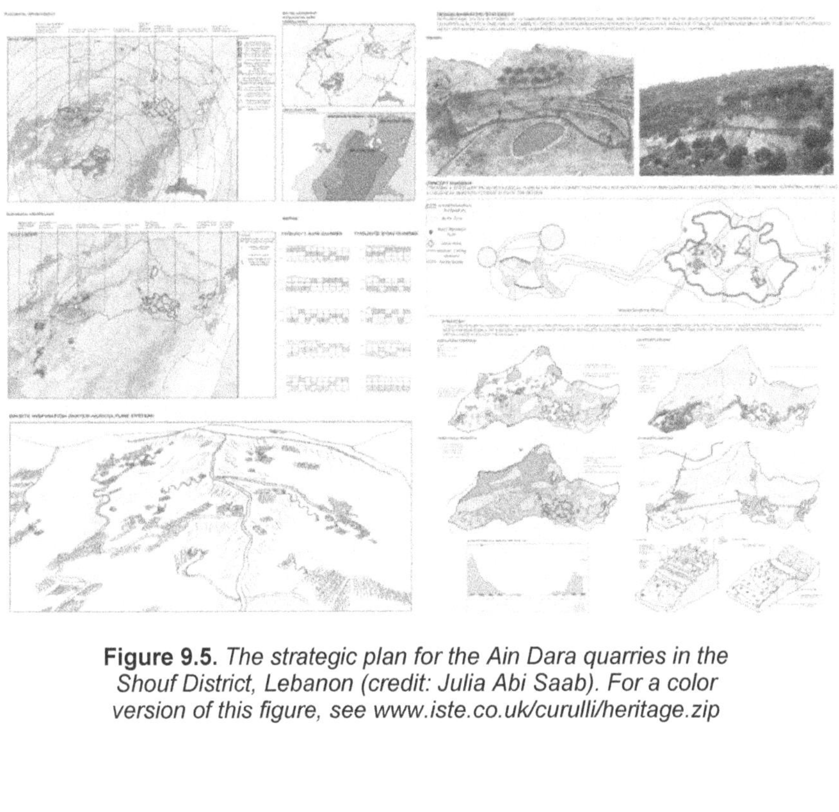

Figure 9.5. *The strategic plan for the Ain Dara quarries in the Shouf District, Lebanon (credit: Julia Abi Saab). For a color version of this figure, see www.iste.co.uk/curulli/heritage.zip*

In particular, they responded to the ecological and environmental needs of the region, reclaiming disturbed ecosystems through constructed wetlands for wastewater treatment, rainwater harvesting, nutrient recovery and quarry re-naturalization. The ecologically rich context in which the quarries are located made the design scenarios responsive to the Natural Reserve, serving as an additional ecological hub to the area (Figure 9.6). The proposal is therefore expected to enhance flora and fauna diversity and create microhabitats within the quarries to link the site to nearby ecological hubs, both ecologically and aesthetically, at a larger scale.

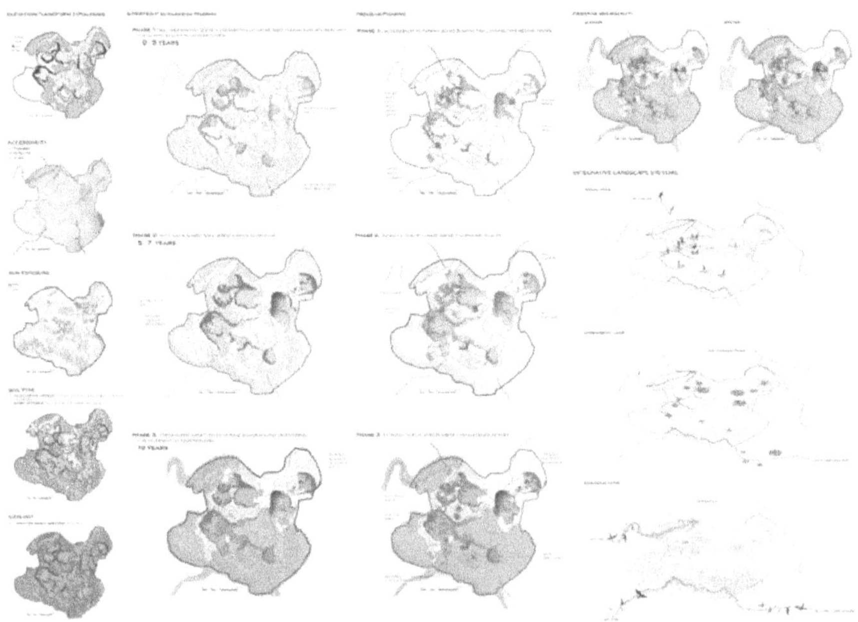

Figure 9.6. *The strategic ecological phasing and the integrative landscape systems for the Ain Dara quarries in the Shouf District, Lebanon (credit: Julia Abi Saab). For a color version of this figure, see www.iste.co.uk/curulli/heritage.zip*

The project is also responsive to culture, while proposing an environment that supports educational and social activities for locals and tourists, which are activities rooted in the Shouf culture (Figure 9.7).

Climate change and population growth pressure affect the availability of water resources, especially in the Middle East. Several studies demonstrated that weather variabilities due to climate change would result in water scarcity in rivers and springs in the Mena region (Bou-Zeid and El-Fadel 2002; Shaban 2011; Bates et al. 2008). The Sayniq River is currently neglected, mistreated and misused due to human-induced pollution from sewage waste, dumping and excessive agricultural activities (Figure 9.8).

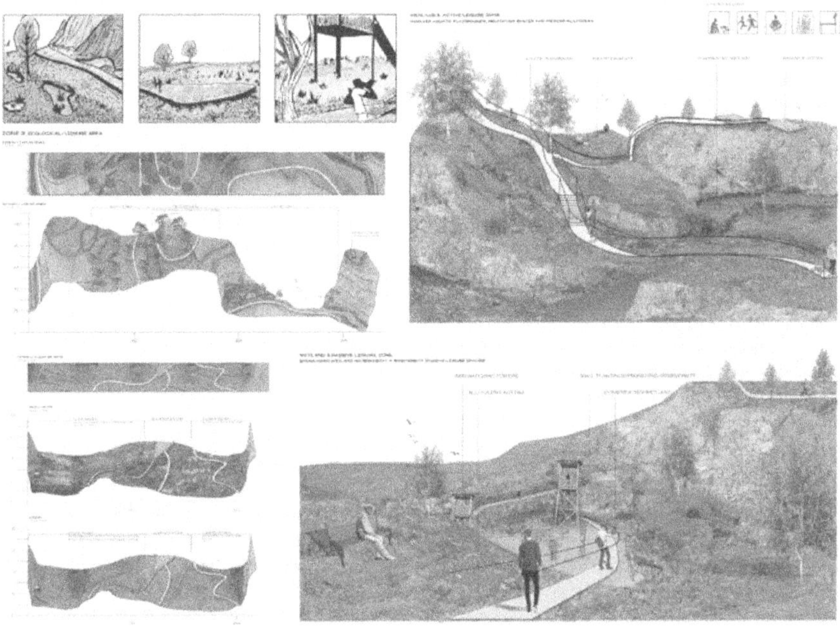

Figure 9.7. *The landscape design of the ecological and leisure practices for the Ain Dara quarries in the Shouf District, Lebanon (credit: Julia Abi Saab). For a color version of this figure, see www.iste.co.uk/curulli/heritage.zip*

The student aimed to transform this disrupted and hidden watercourse into a landscape of ecological and public engagement based on sustainability and social viability practices (Figure 9.9). To this end, naturalized riverbanks, remediated habitats, new accesses to the river edge and redefined public river lines were proposed in the project "Sayniq River: Discovering the Hidden Waterscape". The design strategy focuses on ecological rehabilitation, visual and physical connections improvement, and reclamation of public lands along the watercourse to integrate social hotspots. Ecological connections are proposed by maintaining continuous riparian vegetation on the river shores and enhancing biodiversity through agroforestry practices encompassing remediating flora and re-introducing native forest species within empty niches and agricultural fields.

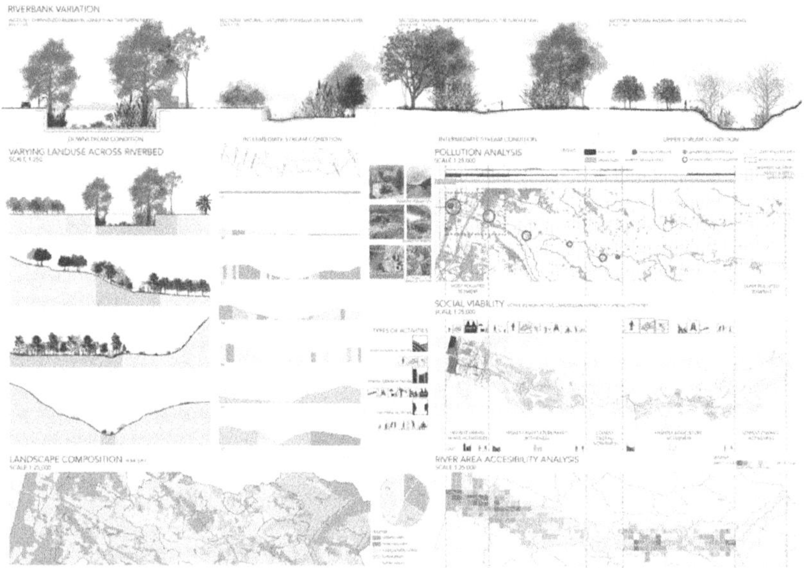

Figure 9.8. *The Sainiq River analysis of the existing conditions, Lebanon (credit: Nour El Zein). For a color version of this figure, see www.iste.co.uk/curulli/heritage.zip*

Rehabilitation is based on native species introduction to achieve a self-sustaining landscape. Several intangible dwelling practices associated with the river (Law and Chen 2020) have been discovered through informal discussions with the locals. Therefore, the project envisioned a revival of social engagement between the community and the river by creating and restructuring trails crossing old mills, churches, open spaces and new leisure places (Figure 9.10). People are brought closer to the river by visually connecting them to its edges. The newly introduced social spaces entail both residents and the visitors of the site (economic, leisure, etc.). The design aims to achieve a landscape of high ecological performance that allows for proper social integration.

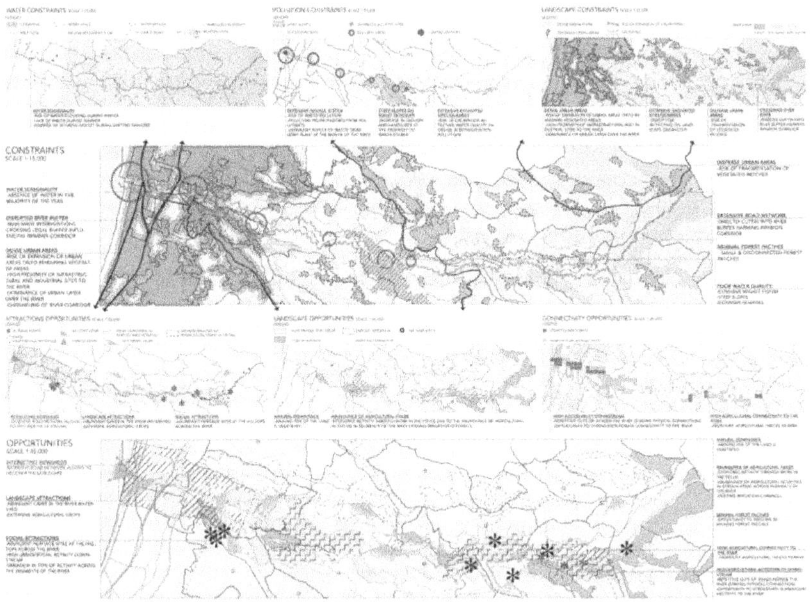

Figure 9.9. *The Sainiq River opportunities and constraints, Lebanon (credit: Nour El Zein). For a color version of this figure, see www.iste.co.uk/curulli/heritage.zip*

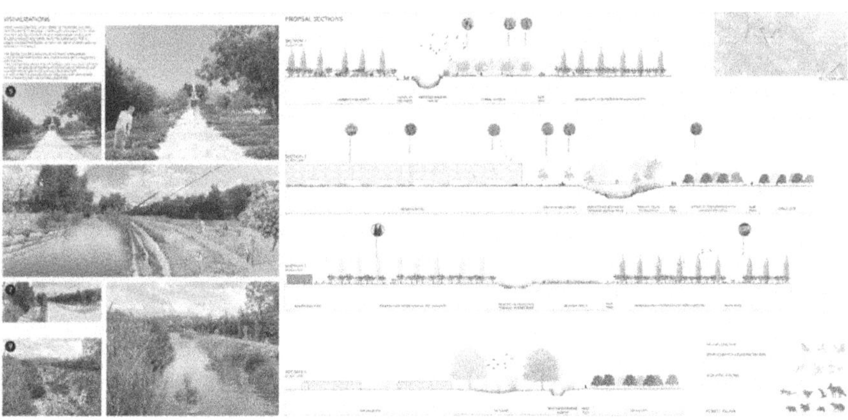

Figure 9.10. *The Sainiq River proposal (credit: Nour El Zein). For a color version of this figure, see www.iste.co.uk/curulli/heritage.zip*

With the project "Reclaiming the Right to Water: Hasbani River" as a resource, the student aims to protect and enhance the river ecosystem to face the future challenges of climate change and enhance the overall microclimate of the surrounding region. Moreover, it reclaimed the right to water, which has been violated since 1950, by proposing sociocultural and environmental recovery scenarios (Figure 9.111). To this end, the project proposed a design framework focused on three major components: treatment and storage, which can ensure the right to clean water for locals in the region, and recreation of the riparian environment, to ensure the social and cultural rights of locals and visitors to the riparian landscape (Figure 9.12).

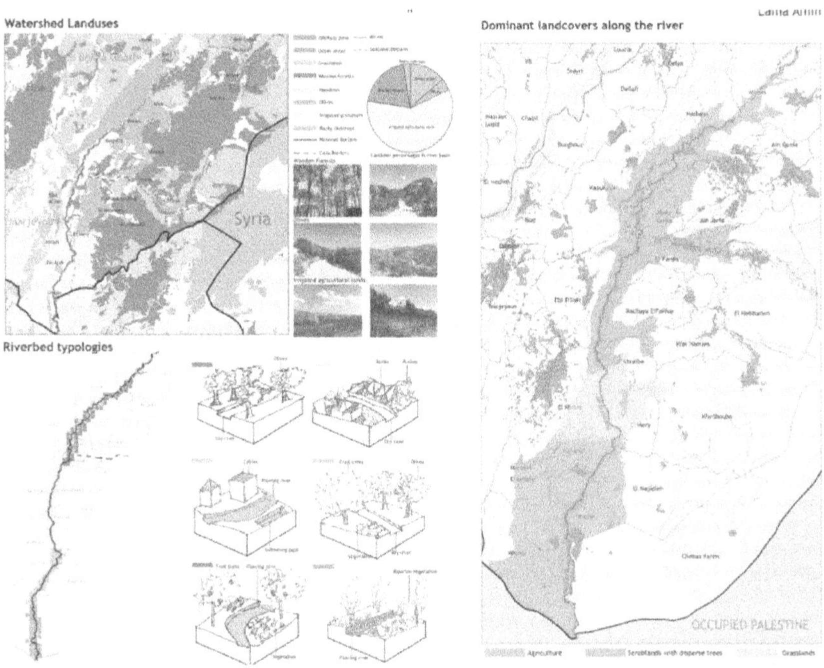

Figure 9.11. *The Hasbani riverbed typologies and dominant landcovers (credit: Lama Amin). For a color version of this figure, see www.iste.co.uk/curulli/heritage.zip*

The water management interventions are proposed at the local level to allow communities to regain and build their new relationship with the water through waterscapes of harvesting, cleaning and accessibility, thus, creating localized models for living in water-stressed areas. Water treatment was planned around the Wazzani Spring, in the southern part of the river, where agricultural activities are prominent.

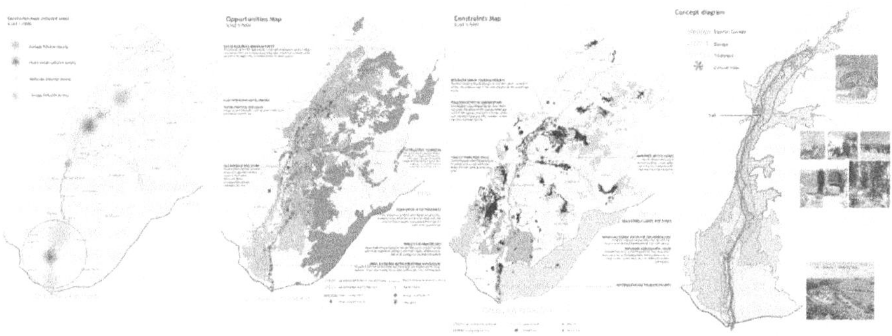

Figure 9.12. *The Hasbani River opportunities and constraints (credit: Lama Amin). For a color version of this figure, see www.iste.co.uk/curulli/heritage.zip*

In this river section, agricultural production is suffering due to the use of polluted water. The middle part of the river, by the villages of ElKhiam and Ebel AlSaki, is almost dry for most of the year. The analysis of the geomorphological character demonstrated that the ground permits water collection. Therefore, storage areas and activities have been planned to respond to the regional request for water (Figure 9.13).

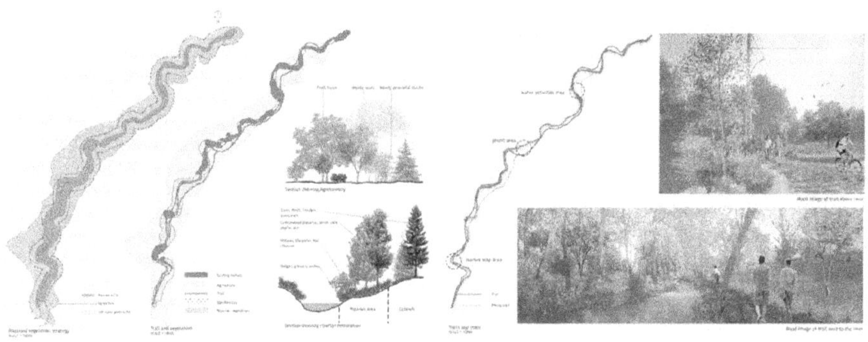

Figure 9.13. *The Hasbani River proposal (credit: Lama Amin). For a color version of this figure, see www.iste.co.uk/curulli/heritage.zip*

9.7. Conclusion

The combined forces of climate change, urban development pressure and socioeconomic and political crisis have intensified the risk on the Lebanese landscapes intended as a testimony of the country's strong heritage. In this part of the world, where the landscape discourse has long been relegated to urban beautification, the experience conducted at the Landscape Capstone Project I and II (LDEM 241-242), during the 2019–2020 academic year, has displayed possible ways to respond to the inevitable threats and consequences of the contemporary weather-related challenges and sociopolitical crisis. While investigating the reasons that could produce alteration, decay, depletion and material and immaterial loss of assets in the landscape, students understood the risks of experimenting and developing design strategies and ideas to protect and restore these landscapes, secure biodiversity and maintain vital ecosystem services, while ensuring shared prosperity and well-being. In their masterplans, green infrastructural networks are the backbones of proposals that sustainably address social, environmental, political and cultural issues resulting from climate challenges and complex regional crises. The proposed landscape approaches to land use considered land and water wildlife habitats, visual resources, heritage vegetation and cultural landscapes as opportunities for social, economic and environmental sustainability. While tackling cultural landscapes at risk, the students proposed comprehensive strategies for multi-faceted needs. Moreover, they used the three pillars of the environmental, social and economic sustainability framework to identify scenarios designed to celebrate the cultural and historical significance of the selected sites.

9.8. Acknowledgments

Many thanks go to the LDEM department at the American University of Beirut, Lebanon, and to the co-teachers of the LDEM 241-242 landscape Capstone project, who highly contributed to the success of the course, Mona Khechen and Balsam El Ariss. A special thanks to all the students of the course. Without them, all of this would not have taken place.

9.9. References

Abou Slaymane, R. and Soliman, M.R. (2022). Integrated water balance and water quality management under future climate change and population growth: A case study of Upper Litani Basin, Lebanon. *Climatic Change*, 172, 28.

Alexander, C., Ishikawa, S., Silverstein, M. (1977). *A Pattern Language: Towns, Buildings, Construction*. Oxford University Press, New York.

Anonymous (2019). Turmoil in Lebanon. *Strategic Comments*, 25(10), iv–vi.

Anonymous (2022). Lebanon's governance crisis. *Strategic Comments*, 28(1), i–iii.

Barthel-Bouchier, D. (2013). *Cultural Heritage and the Challenge of Sustainability*. Walnut Left Coast Press, Creek, CA.

Bates, B., Doell, P., Blanca, J.M. (2008). Climate change and water. Technical Paper of the Intergovernmental Panel on Climate Change, IPCC Secretariat, Geneva, 7(3), R126.

Bou-Zeid, E. and El-Fadel, M. (2002). Climate change and water resources in Lebanon and the Middle East. *Journal of Water Resources Planning and Management*, 128(5), 343–355.

Cacciotti, R., Kaiser, A., Sardella, A., De Nuntiis, P., Drdacky, M., Hanus, C., Bonazza, A. (2021). Climate change-induced disasters and cultural heritage: Optimizing management strategies in Central Europe. *Climate Risk Management*, 32, 100301.

Cassar, M. (2005). *Climate Change and the Historic Environment*. Centre for Sustainable Heritage, London.

Colette, A. (2007). Climate change and World Heritage. Report on predicting and managing the impacts of climate change on World Heritage and Strategy to assist States Parties to implement appropriate management responses. UNESCO World Heritage Centre, Paris.

Colette, A. (2009). *Case Studies on Climate Change and World Heritage*. UNESCO World Heritage Centre, Paris.

Council for Development & Reconstruction (2005). National Physical Master Plan of the Lebanese Territory (NPMPLT). Final report, Lebanon.

Cuca, B. and Agapiou, A. (2018). Impact of land-use change and soil erosion on cultural landscapes: The case of cultural paths and sites in Paphos district, Cyprus. *Appl. Geomat.*, 10, 515–527.

Darwish, T., Faour, G., Khawlie, M. (2004). Assessing soil degradation by landuse-cover change in coastal Lebanon. *Lebanese Science Journal*, 5, 45–59.

Fairclough, G., Herlin, I.S., Swanwick, C. (2018). *Routledge Handbook of Landscape Character Assessment*. Routledge, London.

FAO (2008). AQUASTAT Country profile – Lebanon. Food and Agriculture Organization of the United Nations, Rome.

Guzman, P., Fatorić, S., Ishizawa, M. (2020). Monitoring climate change in World Heritage properties: Evaluating landscape-based approach in the state of conservation system. *Climate*, 8(3), 39.

Harvey, D. and Perry, J. (2015). *The Future of Heritage as Climates Change: Loss, Adaptation and Creativity*. Routledge, Abingdon.

Ivajnšič, D., Kaligarič, M., Fantinato, E., Del Vecchio, S., Buffa, G. (2018). The fate of coastal habitats in the Venice Lagoon from the sea level rise perspective. *Appl. Geogr.*, 98, 34–42.

Johnson, C.E. and Vida, G. (2020). Evaluating the adaptive capacity of cultural landscapes to climate change: Incorporating site-specific knowledge in National Park Service vulnerability assessments. *Parks Stewardship Forum*, 36(1), 49–56.

Jomaa, I., Auda, Y., Abi Saleh, B., Hamze, M., Safi, S. (2008). Landscape spatial dynamics over 38 years under natural and anthropogenic pressures in Mount Lebanon. *Landscape and Urban Planning*, 87, 67–75.

Jørgensen, K., Clemetsen, M., Richardson, T., Halvorsen Thorén, K. (eds) (2016). Introduction. In *Mainstreaming Landscape – A Decade After the European Landscape Convention*. Routledge, New York.

Kaplan, S. (1995). The restorative effects of nature: Toward an integrative framework. *Journal of Environmental Psychology*, 16, 169–182.

Kaplan, R. and Kaplan, S. (1989). *The Experience of Nature: A Psychological Perspective*. Cambridge University Press.

Keellings, D. and Hernández Ayala, J.J. (2019). Extreme rainfall associated with Hurricane Maria over Puerto Rico and its connections to climate variability and change. *Geophys. Res. Lett.*, 46, 2964–2973.

Khater, C. (2004). Dynamiques végétales post perturbation sur les carrières calcaires au Liban. Stratégies pour l'écologie de la restauration en régions méditerranéennes. PhD Thesis, Académie de Montpellier, Université Montpellier II.

Law, A.M. and Chen, X. (2020). "Absent–Present" heritage: The cultural heritage of dwelling on the Changjian (Yangtze) River. In *Adaptive Strategies for Water Heritage*, Hein, C. (ed.). Springer, Cham.

Lelieveld, J., Hadjinicolaou, P., Kostopoulou, E., Chenoweth, J., El Maayar, M., Giannakopoulos, C., Hannides, C., Lange, M., Tanarhte, M., Tyrlis, E. et al. (2012). Climate change and impacts in the Eastern Mediterranean and the Middle East. *Climatic Change*, 114(3), 667–687.

Lemieux, C.J., Beechey, T.J., Gray, P.A. (2011). Prospects for Canada's protected areas in an era of rapid climate change. *Land Use Policy*, 28, 928–941.

Lennard, S.H.C. (1987). *Liveable Cities: People and Places: Social Design Principles for the Future of the City*. Gondolier Press, New York.

Lionello, P. and Scarascia, L. (2018). The relation between climate change in the Mediterranean region and global warming. *Reg. Environ. Change*, 18, 1481–1493.

Lowenthal, D. (2007). Living with and looking at landscape. *Landscape Research*, 32(5), 635–656.

Lynch, K. (1960). *The Image of the City*. The MIT Press, Cambridge, MA.

Lynch, K. (1981). *The Theory of Good City Form*. The MIT Press, Cambridge, MA.

Makhzoumi, J. (2002). Landscape in the Middle East: An inquiry. *Landscape Research*, 27, 213–228.

Makhzoumi, J. (2009). Unfolding landscape in a Lebanese village: Rural heritage in a globalising world. *International Journal of Heritage Studies*, 15(4), 317–337.

Makhzoumi, J. (2016). From urban beautification to a holistic approach: The discourses of "landscape" in the Arab Middle East. *Landscape Research*, 41(4), 461–470.

Makhzoumi, J. and Al-Sabbagh, S. (2018). Landscape and urban governance: Participatory planning of the public realm in Saida, Lebanon. *Land*, 7(2), 48.

Makhzoumi, J., Chmaitelly, H., Lteif, C. (2016). Holistic conservation of bio-cultural diversity in coastal Lebanon: A landscape approach. *Journal of Marine and Island Cultures*, 1, 27–37.

MedECC (2020). Climate and environmental change in the Mediterranean basin – Current situation and risks for the future. In *First Mediterranean Assessment Report*, Cramer, W., Guiot, J., Marini, K. (eds). Union for the Mediterranean, Plan Bleu, UNEP/MAP, Marseille.

Meinig, D.W. (1979). Introduction. In *The Interpretation of Ordinary Landscapes*, Meinig, D.W. (ed.). Oxford University Press, New York.

Ministry of Agriculture (2003). National action programme to combat desertification. Beirut [Online]. Available at: https://knowledge.unccd.int/sites/default/files/naps/lebanon-eng2003.pdf [Accessed January 2023].

Ministry of Environment (MOE) (2011a). Climate change: Vulnerability and adaptation. United Nations Development Programme (UNDP), Beirut.

Ministry of Environment (MOE) (2011b). Lebanon's second national communication to the United Nations Framework Convention on Climate Change. United Nations Development Programme (UNDP), Beirut.

MoE/UNDP/GEF (2019). Lebanon's Third Biennial Update Report (BUR). MoE/UNDP/GEF.

Perry, J. (2011). World Heritage hot spots: A global model identifies the 16 natural heritage properties on the World Heritage list most at risk from climate change. *International Journal of Heritage Studies*, 17(5), 426–441.

Perry, J. (2015). Climate change in the world's best places: A wicked problem in need of immediate attention. *Landscape and Urban Planning*, 133, 1–11.

Rachi, R. (2022). Filthy flows of power: A political ecology of disasters in Lebanon. In *The Lebanon Uprising of 2019. Voices from the Revolution*, Karam, J.G. and Majed, R. (eds). I.B. Tauris, London.

Reimann, L., Vafeidis, A.T., Brown, S., Hinkel, J., Tol, R.S.J. (2018). Mediterranean UNESCO World Heritage at risk from coastal flooding and erosion due to sea-level rise. *Nat. Commun*, 9, 4161.

Selman, P. (2004). Community participation in the planning and management of cultural landscapes. *J. Environ. Plann. Manag.*, 47(3), 365–392.

Shaban, A. (2011). Analysing climatic and hydrologic trends in Lebanon. *Journal of Environmental Science and Engineering*, 11, 343–355.

Snow, K. and Kluiving, S.J. (2022). Reading the Anthropocene through landscape archeology: Historical and contemporary landscapes of accumulation in the Netherlands and Lebanon. *The International Journal of Environmental Sustainability*, 18(2), 1–22.

Stenseke, M. and Jones, M. (2011). Conclusion: Benefits, difficulties, and challenges of participation under the European landscape convention? In *The European Landscape Convention: Challenges of Participation*, Jones, M. and Stenseke, M. (eds). Springer, Dordrecht.

Strydom, S., Murray, K., Wilson, S., Huntley, B., Rule, M., Heithaus, M., Bessey, C., Kendrick, G.A., Burkholder, D., Fraser, M.W., Zdunic, K. (2020). Too hot to handle: Unprecedented seagrass death driven by marine heatwave in a World Heritage area. *Glob. Change Biol.*, 26, 3525–3538.

Taylor, K. (2022). The intellectual context of the cultural landscape construct. In *The Routledge Handbook of Cultural Landscape Heritage in the Asia-Pacific*, 1st edition Silva, K.D., Taylor, K., Jones, D.S. (eds). Routledge, Abingdon.

Trovato, M.G., El-Akl, N., Ali, D., Abou Faker, S. (2021). Syrian refugees in Lebanon: Protection amidst crises. AUB report, HORIZON 2020 – ADMIGOV – Advancing Alternative Migration Governance [Online]. Available at: http://admigov.eu [Accessed January 2023].

Ulrich, R.S. (1983). Aesthetic and affective response to natural environment. In *Human Behaviour and Environment*, Vol. 6, Altman, I. and Wohlwill, J.F. (eds). Plenum Press, New York.

Ulrich, R.S. (1993). Biophilia–biophobia and natural landscapes. In *The Biophilia Hypothesis*, Kellert, R. and Wilson, E.O. (eds). Island Press, Washington, DC.

Ulrich, R.S. (1997). Pre-symposium workshop: A theory of supportive design for health care facilities. *Journal of Health Care Design*, 9, 3–7.

Ulrich, R.S., Simons, R.F., Losito, B.D., Fiorito, E., Miles, M.A., Zelson, M. (1991). Stress recovery during exposure to natural and urban environments. *Journal of Environmental Psychology*, 11, 201–230.

Ward, C. and Ruckstuhl, S. (2017). *Water Scarcity, Climate Change and Conflict in the Middle East: Securing Livelihoods, Building Peace*. Bloomsbury Publishing, London.

Wiener, J. (2017). An enduring fascination with Lebanon: A conversation with Dr. Marielle Martiniani-Reber [Online]. Available at: http://etc.ancient.eu/interviews/an-enduring-fascination-with-lebanon-a-conversation-with-dr-marielle-martiniani-reber/ [Accessed January 2023].

Wilson, E.O. (1993). Biophilia and the conservation ethic. In *The Biophilia Hypothesis*, Kellert, R. and Wilson, E.O. (eds). Island Press, Washington, DC.

10

Cultural Resilience of the Oploo Watermill

Water plays an important role in Dutch history, one of the reasons being its use as an energy source. Watermills generate kinetic energy from the water and have been used to produce food and control the water level in an area. Even after having lost their production function, the watermill buildings have cultural importance. At the same time, they have the potential for climate change mitigation. Watermills have a strong cultural resilience, which is defined as the capability of a building to absorb (economical, functional and sustainable) changes (Holtorf 2018). Furthermore, watermills have a strong presence in local history and culture; they can foster cultural resilience. Watermills are an example of knowledge carriers that can be used for future climate adaptation.

The study discussed in this chapter is part of student research done for the course Built-up Heritage Over Time, provided by the Technical University Eindhoven and taught by Irene Curulli and Deniz Ikiz Kaya. The course focuses on the idea of combining heritage buildings and climate adaptation. Without proper interventions, heritage buildings can become victims of climate change; however, they also provide knowledge on how previous generations tried to address the impact of climate.

During the course, research was carried out on the Brabant watermill landscape, with a focus on the Oploo watermill. The water mill is located in the eastern part of North Brabant in the municipality of Land van Cuijk, at the edge of the village Oploo, surrounded by farmland. This study was focused on analyzing the current

Chapter written by Eva STARK and Lina BEKKER.

use, the history and a social, spatial and architectural impact assessment. This was done in order to gain an overview of the previous role of the mill in the village throughout history, its current role and its potential for the future. Researching this can give some insight into the potential of using the mill to increase the climate resilience of the area.

The research during the course Built-up Heritage Over Time was done in collaboration with Emiel Groenhagen. We want to thank him for his contribution, we could not have done it without him.

10.1. Methodology

The research on the mill was conducted in 2021–2022. Data were collected through an array of methods: literature reviews of municipal sources (historical), books, websites and archives; historical photos found on the internet etc. Technical drawings were supplied by the current users of the mill: Jan van Riet (caretaker of the mill) and Piet van den Oever (Caretaker Guild). Topographic information was based on datasets accessed through the internet; a visit to the site on November 26, 2021, interviews with the current caretakers of the mill and the Guild, an online survey (quantitative) with 108 responses distributed by Jan van Riet to the Guild, the *dorpsraad* (village council), the *Heemkundevereniging* (local history association), and nearby residents via email. The data analysis was done through mapping, facade analysis, site analysis, (historical) photo analysis, (historical) map analysis and digital and physical model.

10.2. History of the Oploo watermill

The history sequence of the watermill was analyzed based on three different topics: historical events that impacted the mill, changes in the watermill landscape and adjustments to the building. The history of the mill can also be viewed in a video[1].

The Oploo watermill was built in 1725 (De Nederlandse Molendatabase, n.d.) and was used as an oil/peeling mill, owned by the *Heren van Steenhuijs* (KvR, n.d.).

1 Watermolen Oploo (1900–2020), directed by Lina Bekker, Emiel Groenhagen and Eva Stark (Eindhoven, 2022). Available at: https://www.youtube.com/watch?v=9tYhsG6p9bg&feature=youtube.

In 1780, an extension was added; it is assumed that around this time the mill started functioning as a pill mill (Verberk and van den Berg-Van Sambeek 2011). Before 1850, the Peel was largely an unexploited, natural high peat area, with high groundwater levels due to the sponge-like working of the peat. Throughout the next century, the peat was excavated and replaced with cultivable soil. As a result, the coverage of the sponge-like substance diminished and groundwater levels dropped (PDOK 2018; Provincie Noord-Brabant n.d.), causing the mill to no longer be able to run during the summertime. It was decided to combat this by building a horse mill and in 1911 the watermill was expanded with a diesel engine (Archieven.nl 2020). In 1932, the mill became a grain mill, and the diesel engine was still used as an energy source during this time (Brabants Historisch Informatie Centrum 2021). During the war, it was an important food source for the inhabitants of Oploo and the surrounding villages (De Nederlandse Molendatabase n.d.). The water level continued to decline until about 1953, after which the natural surface coverage remained largely the same (PDOK 2018; Provincie Noord-Brabant n.d.). Due to the limited time that the mill could rotate, the mechanical system had jammed, making it no longer possible to be used for grain production. The watermill could no longer provide any form of income, causing the building to fall into disrepair (Verberk and van den Berg-Van Sambeek 2011).

The municipality bought the watermill in 1968, and it became a national monument (Rijksdienst voor het Cultureel Erfgoed 2020) and was renovated (Verberk and van den Berg-Van Sambeek 2011). According to the municipality, the mill had to have a social function for the community. In 1990, the mill was rented by the *Peelvlinders* and became a clubhouse; after a few years, it no longer had enough members to exist (Verberk and van den Berg-Van Sambeek 2011). During this time, a second much more expensive renovation took place (Verberk and van den Berg-Van Sambeek 2011).

Its current function is as a connecting place for the villagers of Oploo and is maintained and used by the *Sint Matthias* Guild. Guided tours showing the mill are given with the help of volunteers (Molenaars n.d.). Figure 10.1 shows the timeline of the Oploo watermill.

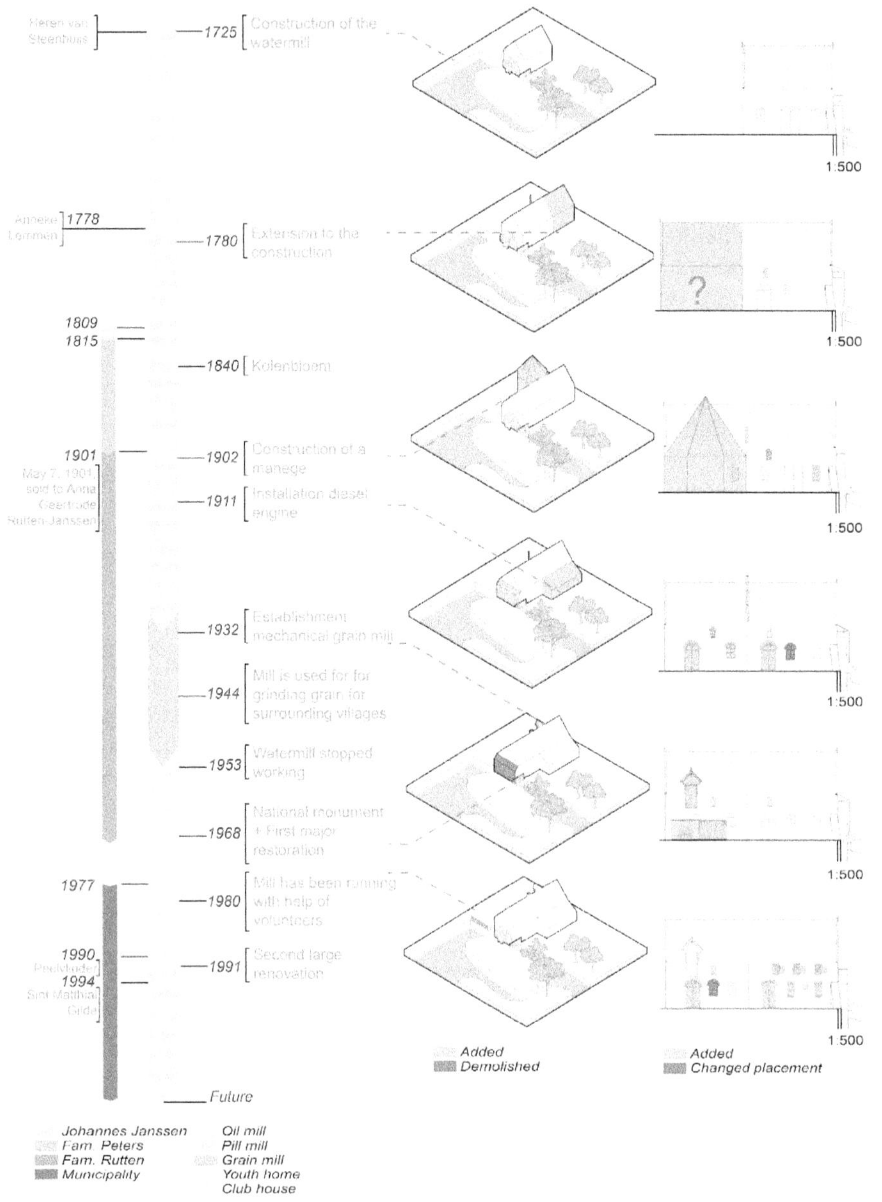

Figure 10.1. *Timeline of the Oploo watermill. For a color version of this figure, see www.iste.co.uk/curulli/heritage.zip*

10.3. The Sint Matthias Guild

The Sint Matthias Guild belongs to a larger network of other guilds, all part of the Noordbrabantse Federatie van Schuttersgilden (Noordbrabantse Federatie van Schuttersgilden n.d.) and has existed since the 16th century (KvR n.d.). However, due to a decrease in members, the Guild was absent for a while. A relaunch was made in 1994 as they took up residence in the Oploo watermill (KvR n.d.).

Even after gaining its monumental status, changes were made to the building to better accommodate its new function; nevertheless, many of the original elements are still visible, as shown in Figure 10.2. Major maintenance is carried out by specialists, which the municipality pays for with a possible subsidy from the National Cultural Heritage Agency (Ministerie van Onderwijs, Cultuur en Wetenschap 2022a). As for maintenance on the inside of the mill, the Guild itself carries out the work; this is paid for by the Guild itself. The Guild gets its income from various sources, one of which is social activities. Without these activities, the Guild could not exist, Covid-19 has been a big stressor on the future of the Guild during 2021–2022.

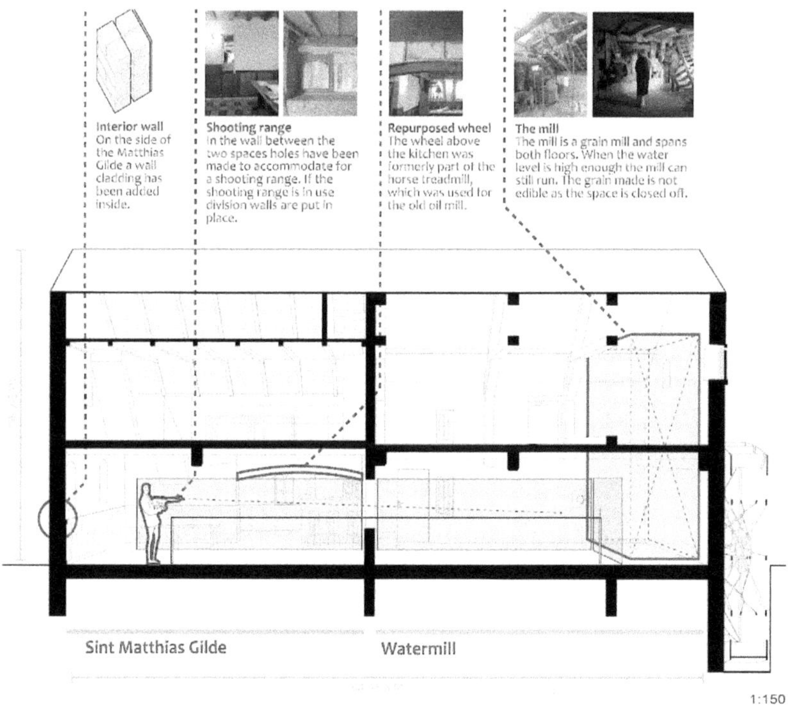

Figure 10.2. *Section (based on drawings provided by Piet van den Oever). For a color version of this figure, see www.iste.co.uk/curulli/heritage.zip*

10.4. Heritage used for climate resilience

Recently, heritage buildings have been experiencing a new threat in the form of climate change. This poses the question: how climate resilient are Dutch dikes, bridges, historical buildings and monuments? Without proper precautions, a chunk of the heritage in the Netherlands will be lost to flooding, droughts or natural disasters (Ministerie van Onderwijs, Cultuur en Wetenschap 2022b). However, it can also offer solutions to climate change. Heritage does not only come in the form of buildings, traditions and knowledge are also part of it. This knowledge is an important element of how the country treats water (Kennisportaal Klimaatadaptie and Fatoric 2021). Heritage offers knowledge on how previous generations limited flooding, but also how they built housing and infrastructure while keeping the landscape in mind (Kennisportaal Klimaatadaptie and Fatoric 2021). It also allows us to learn from the mistakes of previous generations. An example of this is the restoration of watermills and dikes to regulate the water level in an area, therefore regaining their former function (Kennisportaal Klimaatadaptie n.d.). This is already being done to three other watermills in the province of Brabant, the mills located in Venbergen, Opwetten and Spoordonk. The landscape is being readied to be used for water buffering and storage, making it possible to have extra space for water storage in wet periods and water retention in dry periods (Meijer et al. 2020). A difficulty of using heritage buildings to make the country more climate resilient is to determine which buildings should be addressed first. Not every building is impacted the same by climate change, therefore it is required to properly analyze vulnerabilities. However, the funding to address climate change is not divided equally, a national adaptation policy for cultural heritage could help solve this (Kennisportaal Klimaatadaptie and Fatoric 2021).

The impact of climate change is also visible in the watermill in Oploo. Alongside the changes to the function of the mill and building, the landscape changed as well. The Oploosche Molenbeek supplies water to the mill and widens closer to the mill. The water height of the settling pond can be regulated to improve the rotating speed of the mill. The water storage capacity of the watermill landscape has diminished in the past 150 years. Analysis of historical and contemporary maps with the use of GIS software showed that the water supply for the mill flows from De Peel to the Maas. De Peel consists of high-elevation peat grounds (PDOK 2018). In total, the landscape height drops from about 28 m at De Peel to 10 m at the Maas. Around and before 1850, the Peel was largely an unexploited, natural high-peat area with high groundwater levels. Over time, the peat was excavated and replaced with cultivable soil, causing groundwater levels to drop (PDOK 2018; Provincie Noord-Brabant n.d.), as shown in Figure 10.3. To the northwest of the watermill, there is a windmill, the Korenbloem, placed in Oploo in 1845 (Molenaars, n.d.). This was built in times when the water level was too low in the summer and the diesel engine was not yet constructed. Currently, the mill is only able to run during the winter months for showcasing purposes, when the water levels are sufficiently high. The fall of the water at the mill is about 1 m. More recently, in 2010, the village flooded due to an excess of rain (Redactie 2017). Opportunities for fluctuating water level regulation using the mill are present and may be controlled by the local community; these are however not yet exploited. This may be due to a lack of funding, resistance from the local community or a lack of communication between stakeholders.

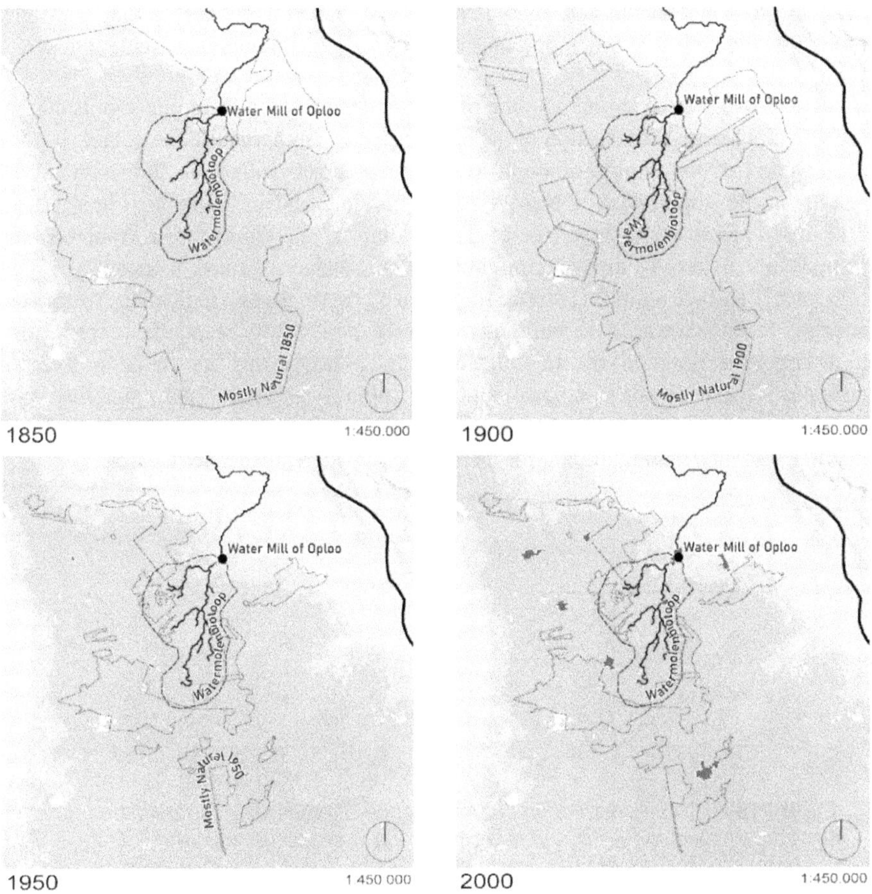

Figure 10.3. *Landscape changes surrounding the Oploo watermill*

10.5. Survey results

To gain a better understanding of the role of the mill in the community, a survey was distributed. The survey was performed by 108 respondents, of which 95% completed the survey fully. The vast majority indicated living in Oploo. Approximately one in three respondents are members of the Guild. The age distribution is nicely balanced, but no respondents indicated to be younger than 20 years old. Because of the Covid-19 pandemic, many respondents indicated that they had not visited the building this last year. The questions asked in the survey ranged from the respondents' knowledge of the current function of the building and how fitting this function is to the importance of preserving the building and its function.

The results show a strong bonding of the respondents to the building and its current use as a clubhouse. Preservation of the building is very important to them, and so is the preservation of the Guild, although to a slightly lesser degree, as shown in survey results on the importance of preserving the building and its function (Figure 10.4). Still, respondents deem the usage of the building by the Guild fitting. When asking respondents to describe the watermill in one word, answers such as "Legendary", "Part of Oploo" and "Cultural Historic Importance" were given, indicating warm and attached feelings towards the building. This may explain why respondents would find it a large deterioration of the village if the building and its function were to disappear. These answers indicate a strong cultural resilience (the capability of a building to absorb economical, functional and sustainable changes (Holtorf 2018)), making it an interesting building for increasing climate resilience in the area.

Figure 10.4. *Survey results on the importance of preserving the building and its function. For a color version of this figure, see www.iste.co.uk/curulli/heritage.zip*

10.6. Conclusion

It is hypothesized that the cultural resilience the watermill fosters comes from the long history of the building, its landscape and the current functioning of the building. Over time, economic and demographic changes affected the landscape in the watermill biotope and the demand for milling products. This, in turn, caused changes in the functionality of the watermill. Utilitarian in nature, the building reflects the functional changes clearly in its facade. As a result, the building appears as a collage that reflects its use over time. Because of the location in the rural landscape and its low economic dynamics, changes in recent decades have been limited, restorative or restricted. The resident survey showed clearly that the watermill serves as a focal point in the local community of Oploo, because of the history of the building, its monumental status, the possibility of recreation on its grounds and near the local water, its use as a clubhouse for the local Guild and as a location for events. It confirmed the warm feelings and feeling of belonging of the water mill to the village of Oploo and the importance of the preservation of the building and the Guild. This highlights the high degree of cultural resilience. However, currently, the municipality is responsible for preserving the monumental building from the outside, while the Guild can alter the interior and make it fit for performing their activities and events that finance it. But simultaneously, the monumental status limits the Guild in making the building fully fit for purpose. High dependence of the Guild on the local community through hosting events may be a long-term problem that in turn might be a municipal problem. Next to this, the Covid-19 pandemic plays a significant role in financial constraints for the Guild. Therefore, for the mill to foster cultural resilience to the highest degree now and in the future, a balance between preserving and altering both building and landscape is important.

An analysis of the past and present of the mill can be useful in dealing with future risks that may impact the watermill. From its history, it becomes clear that the building itself is flexible and can withstand environmental changes; however, since the building gained its monumental status, it has lost some of this flexibility. In the present day, the community has a strong appreciation for the preservation of the mill and its function indicating a new form of flexibility, which is visible in the form of nonpermanent changes that ensure the development of the mill. These capabilities of both the community as well as the building to continue to develop refer to a strong cultural resilience of the watermill. This strong cultural resilience makes the watermill less vulnerable and better able to withstand setbacks. This cultural resilience creates a solid base for local government and policymakers to work with when it comes to making the mill more climate resilient. However, it is important to properly communicate with the community as they are a big part of what makes the watermill capable to handle changes. Heritage buildings come with a lot of opportunities but require being dealt with delicacy.

10.7. References

Archieven.nl (2020). Wilhelmus Rutten, molenaar, Oploo, Sint Anthonis, Ledeacker, het plaatsen van een benzinemotor van 16 pk in een bestaande graanmalerij met watermolen, 11 mei 1911 [Online]. Available at: https://www.archieven.nl/nl/zoeken?mivast=0 [Accessed 22 November, 2021].

Brabants Historisch Informatie Centrum (2021). Rutten, W.J., Oploo v/h B, Oploo: graanmalerij, 4–6-1932 [Online]. Available at: https://www.bhic.nl/integrated?mivast=235 [Accessed 15 December, 2021].

De Nederlandse Molendatabase (n.d.). De Nederlandse Molendatabase (watermolen), Oploo [Online]. Available at: https://molendatabase.nl/nederland/molen.php?nummer=573 [Accessed 19 November, 2021].

Holtorf, C. (2018). Embracing change: How cultural resilience is increased through cultural heritage. *World Archaeology*, 50(4), 639–650.

Jurgens, N. (2008). Rapport over het onderzoek naar de herkomst van molen De Korenbloem in Oploo [Online]. Available at: https://home.hccnet.nl/jefvanveldhoven/27%20Oploo%20Korenbloem%20rapport%20(Molen%20van%20Halder).pdf [Accessed 22 November, 2021].

Kadaster (n.d.). Topotijdreis: 200 jaar topografische kaarten [Online]. Available at: https://www.topotijdreis.nl/ [Accessed 13 December, 2021].

Kennisportaal Klimaatadaptie (n.d.). Kansen van cultureel erfgoed voor klimaatadaptatie [Online]. Available at: https://klimaatadaptatienederland.nl/kennisdossiers/cultureel-erfgoed/kansen-cultureel-erfgoed-klimaatadaptatie/ [Accessed 5 January, 2023].

Kennisportaal Klimaatadaptie and Fatoric (2021). Als je erfgoed verliest, is het voor altijd weg [Online]. Available at: https://klimaatadaptatienederland.nl/actueel/actueel/interviews/cultureel-erfgoed/ [Accessed 5 January, 2023].

Kring van Schuttersgilden Land van Cuijk (n.d.). Kring van Schuttersgilden Land van Cuijk – St. Matthias Gilde Oploo [Online]. Available at: https://www.kringlandvancuijk.nl/index.php/aangesloten-gilden/gilde-oploo [Accessed 13 January, 2022].

KvR. (n.d.). Watermolen en accommodatie [Online]. Available at: https://www.matthiasgilde.nl/watermolen.html [Accessed 22 November, 2021].

Meijer, R., Haarmann, W., Elemans, H., de Mars, H., Bleumink, H. (2020). Watermolenlandschappen voor Klimaatadaptatie: Plan van Aanpak Erfgoed Deal-project voor het Stroomgebied de Dommel [Online]. Available at: https://www.watermolenlandschappen.nl/.

Ministerie van Onderwijs, Cultuur en Wetenschap (2022a). Subsidie instandhouding rijksmonumenten [Online]. Available at: https://www.cultureelerfgoed.nl/onderwerpen/subsidie-instandhouding-rijksmonumenten [Accessed 5 January, 2023].

Ministerie van Onderwijs, Cultuur en Wetenschap (2022b). Klimaatverandering en erfgoed [Online]. Available at: https://www.cultureelerfgoed.nl/onderwerpen/water-en-klimaat/klimaatverandering-en-erfgoed [Accessed 5 January, 2023].

Molenaars, V. (n.d.). De Korenbloem [Online]. Available at: https://dekorenbloemoploo.nl/molenaars [Accessed 21 November, 2021].

Noordbrabantse Federatie van Schuttersgilden (n.d.). De Kringen [Online]. Available at: https://www.schuttersgilden.nl/kringen.html [Accessed 13 January, 2022].

PDOK (2018). Actueel Hoogtebestand Nederland 3 WMS [Online]. Available at: https://www.nationaalgeoregister.nl/geonetwork/srv/api/records/94e5b115-bece-4140-99ed-93b8f363948e.

Provincie Noord-Brabant (n.d.). Noord-Brabant maps [Online]. Available at: https://noord-brabant.maps.arcgis.com/apps/MapJournal/index.html?appid=ea3028b39d927e39b78550d8 [Accessed 13 January, 2022].

Provincie Noord-Brabant (2006). Hoofdwaterlopen 2005 [Online]. Available at: https://www.nationaalgeoregister.nl/geonetwork/srv/api/records/ebc8ef0e-41d5-44d2-b41d-334dd81bf262.

Redactie (2017). Heemschuur Oploo onder water [Online]. Available at: https://www.gelderlander.nl/maasland/heemschuur-oploo-onder-water~aaa87f8a/ [Accessed 25 January, 2022].

Rijksdienst voor het Cultureel Erfgoed (2020). 31772 Watermolenstraat 2 5841 CT te Oploo [Online]. Available at: https://monumentenregister.cultureelerfgoed.nl/monumenten/31772 [Accessed 19 November, 2021].

Verberk and van den Berg-Van Sambeek (2011). Alle Minse van Oploo, Deel 6. 1st ed. Fortrass-Boxmeer. Watermolen van Oploo [Online]. Available at: http://www.waterradmolens.nl/NrdBrabant/Oploo.html [Accessed 23 November, 2021].

Watermolenlandschappen voor Klimaatadaptatie (n.d.). [Online]. Available at: https://www.watermolenlandschappen.nl/ [Accessed 5 January, 2023].

PART 3

Education on Climate Adaptation of Heritage

11

Case-Based Teaching for Climate Heritage Education: Climate Adaptation of Historic Watermill Landscapes as a Case Study

This chapter demonstrates how heritage education connected to climate science and climate action is taught in formative education where students are guided to develop a holistic perspective toward the understanding of cultural heritage, its values, vulnerabilities and adaptive capacity to tackle the impacts of climate change through case studies and stakeholder engagement. It presents two separate educational activities concentrating on climate adaptation of historic watermill landscapes in the Netherlands where different pedagogies are employed in a multi-disciplinary learning environment addressing higher education students from diverse disciplinary backgrounds and with different skill sets. In both cases, the students experiment with design products in climate-related heritage pedagogies, which enabled explorations toward the integration of interdisciplinary knowledge with real-life cases and actors where solutions are illustrated in design outputs.

11.1. Introduction

Cultural heritage plays a significant role in the achievement of sustainable development goals and climate action (ICOMOS 2019). Yet, the core focus areas of heritage education encounter challenges in adapting heritage education and professional training activities to cope with these efforts and address the increasing impacts of climate change. It is important to employ an interdisciplinary and holistic

---

Chapter written by Deniz IKIZ KAYA and Irene G. CURULLI.

approach toward the contextualization of cultural heritage, and how it can act as a resource and vehicle for climate action. The heritage education curriculum should thus stress the vulnerability of built heritage to climate change, as well as its role in climate mitigation and adaptation efforts. This requires innovative pedagogies and teaching/learning methods that foster high levels of stakeholder engagements and interactions, support interdisciplinary approaches to embrace the complexity of climate change and emphasize a critical mindset.

Concerning this existing gap in heritage education, this chapter aims to demonstrate how heritage education connected to climate science and climate action is taught in formative education where students are guided to develop a holistic perspective toward the understanding of cultural heritage, its values, vulnerabilities and adaptive capacity to tackle the impacts of climate change through case studies and stakeholder engagement. The inclusion of case studies within heritage education that deals with the built heritage, and the built environment in general, engages the students in real-life practices, helps them better understand the climatic impacts on built structures and landscapes and how they change over time to adapt, and assess the sociocultural, economic and environmental impacts associated with this change. In addition, connecting students to diverse stakeholders through case studies also equips them with soft skills that better prepare them for the complex decision-making that requires finding a balance between physical interventions, values and interests of various stakeholder groups, including local communities, and the planetary boundaries.

The core elements of this chapter arise from two separate educational activities that concentrated on similar case studies but involved higher education students from diverse disciplinary backgrounds and with different skill sets. Integrating knowledge of heritage and climate science with design thinking, different pedagogies were employed and different outputs were produced as an outcome of these two education activities. The first example is a graduate course targeting architecture and urbanism students, in which they conducted a retrospective heritage impact assessment examining historic structures and cultural landscapes that are adaptively reused, in regard to their heritage significance and how they are integrated into the wider urban development and climate action agenda. The second example is an international student workshop bringing together graduate-level students from heritage, tourism, architecture and engineering studies where they worked on climate adaptation solutions for the same type of heritage landscapes. The different teaching/learning methods employed in these examples through case study analysis are further explained and a discussion is held regarding pedagogical approaches for climate heritage education.

11.2. Case study

As part of these two educational activities concentrating on the adaptation of cultural heritage to the impacts of climate change, the case study teaching method is employed. Case-based mode of instruction is used to stimulate real-life situations where students are asked to actively engage with complex problems and they are encouraged to learn by doing (Anderson and Schiano 2014). As they work through the case with a problem to be solved, they foster their problem-solving and decision-making skills, as well as communication and collaborative skills as they engage with different parties (Krain 2016).

In these two examples, the selected case studies are historic watermill landscapes located in the North Brabant region of the Netherlands. Historic hydropower sites including watermills are cultural heritage assets with a great potential for small hydropower energy production. According to a recent Europe-wide study, the potential contribution of historic watermills to small hydropower potential is estimated to be 16.5% in the EU countries (Punysy et al. 2019). In addition to potential renewable energy production, watermills also possess multiple cultural and economic values with high non-energy benefits. These assets can be reused for multiple purposes, contributing to the regeneration of their environment and providing local economic revenue. Historic watermill landscapes could also be used to tackle climate change impacts: in wet periods, watermill landscapes can contribute to the temporary storage of water and prevent flooding. During summertime, they can reduce aridification and heat stress in urban areas and support agriculture.

Yet, the adaptive capacity of watermill heritage and its potential benefits to climate change adaptation are barely mentioned in the scientific literature, as well as in national adaptation policies and actions. There are currently 150 watermills still existing in the Netherlands. The government's approach to watermills and their surrounding landscapes is mainly based on their heritage value as monuments, their place in the landscape and their function in society (Rijkdienst voor het Cultureel Erfgoed (RCE) 2012). Their economic, environmental and energy-related benefits are so far underestimated or neglected.

In the existence of such knowledge and policy-related gaps, the Province of North Brabant, along with several regional and local partners including the North Brabant Mill Foundation and the Dommel Water Authority, initiated a nationally funded heritage project that concentrated on the values and adaptive potential of historic watermill landscapes for climate adaptation. Collaborating with these partners, the watermill landscapes examined as part of this project are also introduced as case studies for the two educational activities (Figure 11.1).

Figure 11.1. *Fieldtrip to Opwetten watermill landscape. For a color version of this figure, see www.iste.co.uk/curulli/heritage.zip*

11.3. The educational activities

The first educational activity was part of a graduate course designed for the students of the Master Program in Architecture, Building and Planning of the Eindhoven University of Technology in the Netherlands. The goal of this course is to expand the knowledge of built environment students on built heritage, its adaptive reuse and transformation, and to improve their problem-solving, critical thinking and collaboration skills through working on specific heritage cases that have been adaptively reused and that are vulnerable to climatic impacts.

As part of the course, students in groups conduct risk and heritage impact assessments to analyze their development and transformation processes over time, to identify their current state of conservation and vulnerabilities and to assess the physical, environmental, social, cultural and economic impacts of their reuse. In this context, students conduct research in three steps to examine the past, present and future of historic structures and landscapes: (i) pre-design, which is an in-depth study of their historical development and change over time; (ii) design, which focuses on their current state and their reuse process; and (iii) impact assessment, which analyses the results and impacts of the reuse practice. For this purpose, they employ a number of research methods to address each objective, which include historical and archival research, site observation and spatial analysis, interviews, surveys, stakeholder analysis and social media content analysis. This type of case-based learning mode allows them to deal with a real-life site and its challenges, engage with relevant stakeholders (i.e. property owners, managers, local authorities and visitors) and community groups, and equip them with relevant research and soft skills.

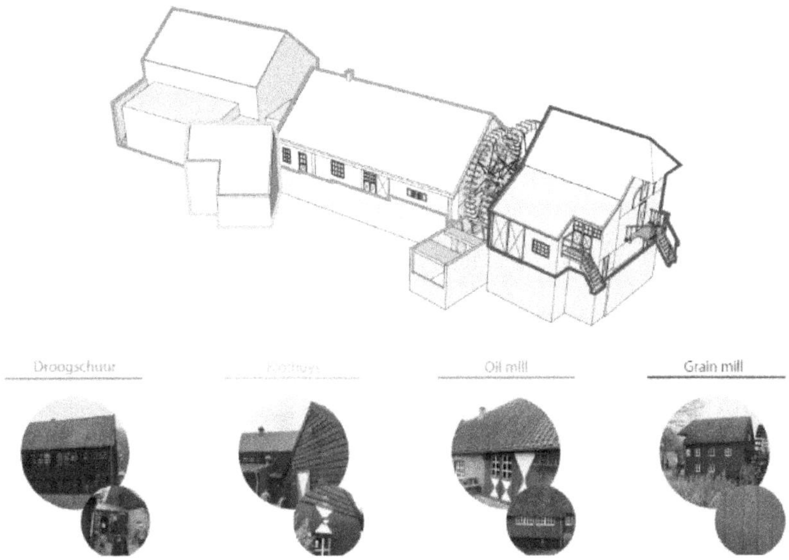

Figure 11.2. *The state of conservation and functional analysis of the Opwetten watermill (work by I. Conti, V. van't Klooster, M. de Boer, 2021). For a color version of this figure, see www.iste.co.uk/curulli/heritage.zip*

An example is the assessment of the Opwetten watermill landscape by a group of students, which was also one of the case areas evaluated as part of the nationally funded collaborative research project. The Opwetten watermill was built alongside the Kleine Dommel, which was originally used for oil, flour, leather, wool and wood production (Figure 11.2). In 2011, it was transformed into a mixed-use complex including a still functioning grain mill, the miller's house, an events space and a restaurant. The watermill and the surrounding landscape have changed significantly over time, from wetlands and agricultural lands to industrial spaces and residential areas surrounded by natural reserves, and the watermill has continued playing a significant role in water management. The renovation and new function of the Opwetten mill mainly had positive socioeconomic impacts on the area, creating a new local economic revenue, but the increasing climate impacts (dry summers, uncontrolled flooding) combined with changes in land use planning and ownership, threaten the integrity of the site and hinder the climate adaptation process.

Figure 11.3. *Design proposal for the reconstruction of a former watermill building and its catchment area with an artificial island (work by B. Mutlu, U Doganay, N. Guglielmi, C.A. Chatzopoulou, Y. Boyanova, Z. Long, G. Maillard, 2021). For a color version of this figure, see www.iste.co.uk/curulli/heritage.zip*

The second educational activity is a student workshop organized as part of the EU-funded education for Climate Resilient European Heritage Architecture (e-CREHA) Erasmus+ project. The workshop aimed at designing for a climate-resilient cultural landscape and took place in Eindhoven in August 2021 (Curulli and Ikiz Kaya 2021). As a case study, the students again focused on the watermill heritage landscapes and collaborated with the local and regional authorities involved in the area. Students of different disciplines, such as architecture, geography, heritage and tourism studies, and computer sciences joined from all over Europe. They came together with local stakeholders, including the municipalities, regional actors, water and landscape bodies, heritage experts and community groups to discuss and design for a climate-resilient future for a cultural landscape area with water heritage assets. As a first step, they identified three different spatial areas to concentrate on and assessed their current state and characteristics. Then, together with the local stakeholders, they worked on several future scenarios for these areas and developed certain design proposals. These design proposals aimed to adapt the landscapes to the impacts of climate change. For instance, they proposed the reconstruction of a former watermill building and its catchment area with an artificial island that could be flooded in wet seasons to mitigate flooding and rising water levels (Figure 11.3).

These two education activities demonstrate case-based, design-oriented pedagogies and teaching methods using built heritage as a tool for climate and heritage education. In both examples, students and young professionals from applied sciences, such as engineers, architects and planners, examined watermill landscapes as case studies, which helped them to develop a holistic interdisciplinary perspective and an inclusive approach to heritage education for climate adaptation. Through stakeholder engagement and challenge-based teaching, they also learned by doing through real-life practices and experiences, which prepare them for complex decision-making in professional life.

11.4. Discussion: design-oriented pedagogical approaches for climate heritage education

A recent EU report that discusses the meeting points of the European Green Deal with cultural heritage indicates that the connection between cultural heritage and climate change has been barely addressed in any education system, including higher education, and urges the need for integrating climate change education with cultural heritage, and vice versa (European Commission 2022). The planning and architectural curriculum mainly concentrates on the built heritage, thus often neglecting the socioeconomic, cultural and intangible attributes associated with it. In addition, the significant role that heritage can play in promoting climate action and its support for efforts in climate change mitigation and adaptation are not addressed in formal heritage and spatial education (Messenger and Bender 2019). It is thus highly important to implement a holistic approach toward the teaching of heritage, even for the applied sciences and engineering, and address the vulnerabilities of the built heritage to climate change, as well as the role of heritage in the development of climate adaptation and mitigation strategies.

In the pathway to achieving a more holistic and interdisciplinary heritage education, innovative pedagogies that accommodate multidisciplinary knowledge and integrated and inclusive teaching and learning activities that use heritage as a catalyst for climate action are vital. In this context, the two educational activities introduced here act as pedagogical experiments in heritage education that can transform its teaching toward a new paradigm, which is needed for driving sustainability and climate action. Both of these educational activities employ a case-based mode of teaching where students undertake a complex challenge embedded in a real-life site, and actively engage with local stakeholders and communities in the process. Following a learning-by-doing method, the students experience directly the broader context and significance of heritage and are encouraged to assess critically its values, vulnerabilities and adaptive capacity with regard to climate change.

In both examples, design outputs are produced as a tool to demonstrate the pedagogical content even though not all of the involved students had prior design education or training. Traditional design education heavily relies on project work and is conducted in design studios, in which product orientation is the main target. (Lawson and Dorst 2009). Experimenting with design products in climate-related heritage pedagogies enabled explorations toward the integration of interdisciplinary knowledge with real-life cases and actors where solutions are illustrated in design outputs. Deploying an interdisciplinary approach and employing varied tools derived from multiple pedagogies can contribute to the transformation of heritage education, which can then address the current global challenges in a more inclusive and timely manner.

11.5. Acknowledgments

The authors would like to thank the TU/e Master in Architecture, Building and Planning students who registered for the Built-up Heritage over Time course in 2021, and to all the organizers, tutors, participants and contributors involved in the Climate Xtremes and Resilient Heritage International Workshop that took place in Eindhoven on August 30–September 8, 2021. This workshop was funded by the education for Climate Resilient European Heritage Architecture (e-CREHA) European Commission Erasmus+ KA203 project (N.o. 2020-1-NL01-KA203-064610).

11.6. References

Andersen, E. and Schiano, B. (2014). *Teaching with Cases: A Practical Guide.* Harvard Business Press, Brighton, MA.

Curulli, I. and Ikiz Kaya, D. (2021). Designing for a climate resilient landscape: ISP1: International workshop on climate xtremes and resilient heritage. Thesis, Technische Universiteit Eindhoven.

European Commission (2022). *Strengthening Cultural Heritage Resilience for Climate Change: Where the European Green Deal Meets Cultural Heritage.* Publications Office of the European Union, Luxembourg.

ICOMOS Climate Change and Cultural Heritage Working Group (2019). *The Future of Our Pasts: Engaging Cultural Heritage in Climate Action.* ICOMOS, Paris.

Krain, M. (2016). Putting the learning in case learning? The effects of case-based approaches on student knowledge, attitudes, and engagement. *Journal on Excellence in College Teaching*, 27(2), 131–153.

Lawson, B. and Dorst, K. (2009). *Design Expertise.* Routledge, London.

Messenger, P.M. and Bender, S.J. (2019). *History and Approaches to Heritage Studies.* University Press of Florida, Gainesville, FL.

Punys, P., Kvaraciejus, A., Dumbrauskas, A., Silinis, L., Popa, B. (2019). An assessment of micro-hydropower potential at historic watermill, weir, and non-powered dam sites in selected EU countries. *Renewable Energy*, 133, 1108–1123.

Rijkdienst voor het Cultureel Erfgoed (RCE) (2012). A future for mills: Principles for dealing with heritage mills. Document, Dutch Cultural Heritage Agency.

12

Increasing Climate Change Resilience by Adapting 20th Century Built Heritage

Critical reviews of many studies on climate change and cultural heritage argue that one of the most prominent problems in the field is the lack of adaptation strategies for increasing the resilience of heritage assets. Conservation, management, impact and risk assessment of archaeological sites and traditional settlements draw attention as the importance of local knowledge and practices are recognized for developing resilience. However, 20th century architecture is left out of these discussions and requires a different type of extensive research in heritage education. Since modern architecture is flexible, durable and free from regional borders and national identities, it can easily be the foundation for studies of international collaboration against climate risks. In this context, the answer to increasing resilience can be found in adapting 20th century heritage within its own framework. Therefore, the study aims to draw attention to the adaptation of 20th century built heritage to boost resilience against climate change by creating a framework and emphasizing its inclusion in heritage education. The methodology for this cause includes determining the existing values of the 20th century built heritage and developing corresponding adaptation and conservation principles within the consideration of local needs and demands. Due to their technological advancement, modern materials like concrete and steel can enable the production of any desired space or installation in terms of geometry and size, as the interiors of these structures can span large distances. Thus, they become suitable for adaptations through interventions involving reuse with additions and removals that the traditional structures cannot accommodate. The advantages arising from these features should be discussed more to broaden the perspectives on heritage education.

Chapter written by Nihan BULUT.

12.1. Introduction

12.1.1. *Problem definition*

Climate change is the biggest threat affecting all kinds of cultural heritage. Therefore, people and their cultures have been the focus of this crisis. It has become crucial to emphasize the urgency of the situation and bring about collective climate action with the participation of all actors (ICOMOS 2021, p. 4).

Since climate change has already become unavoidable, global action and collaboration are needed to develop resilience, and cultural heritage is essential for generating a solution. The resilience of cultural heritage may be understood as the ability to stand against adverse effects, while preserving the existing values. However, it could also be the ability to advance despite the challenging conditions. Therefore, experts agree that it is necessary to determine the contributing climatic factors and their risks to understand the possible threats to heritage (Bosher et al. 2020; ICOMOS 2021). Many known hazards to cultural heritage include environmental disasters such as earthquakes, landslides, fires, floods, rising sea levels, strong winds, freeze-thaw cycles, volcanic eruptions and lightning strikes. While climate change may not directly cause some of these problems, it is arguably a contributing factor (Bosher et al. 2020; Romão and Bertolin 2022).

When the literature is examined, many studies on risk and impact assessment of cultural heritage emerge, but adaptation studies take a much smaller share and need more research (Fatorić and Seekamp 2017a; Orr et al. 2021; Sesana et al. 2021). The reasons for this situation include the lack of guidelines for planning and implementation of actions, insufficient funding, limited information and limited cooperation between different actors (Fatorić and Seekamp 2017b). Although the examples for adaptation are fewer, the information obtained from the studies on traditional knowledge and local practices has benefited the measures to be taken against climate change for heritage sites (ICOMOS 2019).

However, when it comes to 20th century built heritage, the physical aspects of traditional knowledge and practices may not be relevant or even applicable since the modern solutions were thought to be international. On the other hand, the perception and interpretation of these modern structures are authentic to their regions and communities (Bozdoğan 2002). For this reason, modern heritage structures are not as international as they were first thought to be. They can gain new identities according to their local context and acquire new values that can be useful in developing different perspectives.

12.1.2. *Research aim*

Considering this understanding, the research aims to create a framework to show that resilience can be generated against the environmental problems caused by climate change, while conserving 20th century built heritage. At the same time, it acknowledges that these structures contributed to the creation of these environmental problems in the first place. Thus, to develop resilience and conserve these structures, the research strategy is to utilize an adaptation framework for the reuse of modern built heritage in sustainable ways and regard them as existing building stocks for implementing contemporary installations for various purposes that answer environmental needs.

Climate change is affecting our built environments at an alarming pace. To act against it, generating resilience is necessary to slow this process and reduce some contributing factors. Decreasing the production of waste buildings and the number of constructions depends on reevaluating the existing building stock (Plevoets and Van Cleempoel 2019). Meanwhile, all kinds of cultural heritage sites require protection against environmental threats, but 20th century built heritage of modern forms and concrete mostly stays in the shadows of the studies on traditional and archaeological sites. Therefore, when these two challenges are overlapped, the adaptation of 20th century built heritage can be utilized to reduce the pace of climate change and provide awareness for their conservation.

Moreover, heritage education is crucial for initiating actions that promote an understanding of ideas and values in heritage, including its interpretation. It can be a tool to empower and engage with more people, especially the younger generation (European Commission 2019). Thus, it is thought that these structures should be prioritized on the platform of heritage education, since the standard materials and techniques can provide ease in their adaptation and lead to impactful results.

12.2. Adaptation of 20th century built heritage and its methodology

12.2.1. *Understanding the context*

In the past few decades, the developments of our societies have increased the pressure on cultural heritage and resources. While sustainable development is a global priority in the age of climate change, the need for energy saving, energy efficiency and transition to renewable energy has also increased. Retrofitting and refurbishment of historic structures have become important discussions. Maintaining a balance between conservation and improvements through modern technology and materials emerged as a design challenge. Monitoring and measurement of the findings become critical practices (Hassan and Xie 2020).

When modernism emerged at the turn of the 20th century, it was groundbreaking as a new approach to living standards, architecture and construction. Modern buildings differed from existing styles as they challenged outdated living conditions and spatial configurations. They offered flexible and durable structures that utilized large spanning pure geometric forms because of the use of new construction techniques and materials such as reinforced concrete and steel. Thus, they proved ease in the rapidly emerging urban developments and were favored by many architects and designers as a reaction to the constraints of other styles (Le Corbusier 2008 (originally published in 1965)).

Although their architectural aspects were stripped of any regional or national identity as modern solutions to functional and structural concerns were international, their symbolic meanings and the perceptions of the community they serve differed from each other (Bozdoğan 2002). Thus, they soon became authentic cultural assets that enabled them to be viewed as unique heritage structures.

Meanwhile, since the 1800s and the emergence of industrial developments, long-term shifts in temperatures and climatic conditions have occurred mainly because of human activities involving non-renewable energy resources. The consequences of climate change include, but are not limited to, intense droughts, water scarcity, severe fires, rising sea levels, flooding, melting polar ice, catastrophic storms and declining biodiversity. At the same time, the primary sources of emissions due to climate change are energy, industry, transportation, agriculture, land use and, inevitably, buildings. Therefore, global frameworks and initiatives such as the Paris Agreement, UN Framework Convention on Climate Change and Sustainable Development Goals aim to lead the progress against climate change impacts through three main types of action, which are reducing emissions, adapting to climate impacts and funding these actions (United Nations 2022).

Adaptation as a key action requires tools for generation. In that regard, ICOMOS states that heritage can be an asset for adaptation and that cultural heritage can support adaptative measures when its values are incorporated into adaptation governance (ICOMOS 2019; García-Esparza 2019). Therefore, the methodology for this cause needs to start with the determination of the existing conditions and values of the buildings and then develop corresponding adaptation and conservation decisions that involve suitable proposals in consideration of local values, needs and demands. For the adaptation process, information on values, integrity, authenticity and resources are all required (Sesana et al. 2018).

12.2.2. *Significance of the 20th century built heritage for adaptation*

The question of why 20th century built heritage is significant becomes crucial in this discussion. Climate change mitigation goals have put pressure on reducing the carbon footprint of cultural heritage buildings. Measures widely adopted to reduce greenhouse gas emissions of historic buildings aim to insulate the building envelope and improve energy efficiency by renewing their heating, cooling and lighting systems. Complex problems come with these, such as preserving authenticity, maintaining traditional behavior and preventing the decay processes. Therefore, solutions should be selected if they are suitable for the properties of heritage materials (Sesana et al. 2019). However, modern heritage can accommodate these systems more comprehensively based on their character. Reinforced concrete and steel can enable the production of any desired space in terms of geometry and size, while their interiors can span large distances due to their technological advancement. Therefore, they become suitable candidates for adaptations through interventions involving additions or removals that the traditional structures cannot accommodate.

Another difference between traditional cultural heritage and modern heritage is that the latter may appear as secondary in consideration of climatic risks and hazards. But due to their extensive number in city centers, relatively up-to-date construction and large spaces with high load-bearing capacity, they emerge as ideal structures that are widely used around the world. Therefore, they become suitable candidates for the insertion of new components. The additions can be carried by or carry the original structure or enable the new function by providing necessary equipment. Note that 20th century built heritage, including industrial facilities, offer structurally and functionally flexible shells. They are made of durable materials to accommodate demanding functions that last long periods. Additionally, they are flexible enough to adapt to various purposes as a container. If such an approach were to be taken, the adaptation process could lead to a reuse proposal (Yoney et al. 2016).

In the meantime, the same climate hazards affect 20th century heritage, especially concrete and steel as building materials. Frost damage from freezing and thawing cycles and corrosion due to the increase in CO_2 gases are two of the biggest threats to these materials. To overcome climate-related threats, both technological and nature-based solutions can be researched. There could be proposals where the issue of flooding could be solved by either elevating the buildings by employing technology or integrating green spaces that increase permeability through a nature-based approach (Brandt and Rouillard 2020). In this regard, any approach can be considered since adaptation, as a term, refers to a process of fitting. One of the practices particularly focuses on the change in usage. Adaptive reuse is an approach in which the existing building – in this case, a historical heritage – can be repurposed to serve another function. However, this is a complex situation due to many factors contributing to the heritage sites, including significance, various values and stakeholders. There are also no definite steps of action or rules, but it is critical to assess the process from pre-project to post-project phases (Hedieh Arfa et al. 2022).

In light of these, the reason why the 20th century built heritage is suitable for adaptation despite its neglect can be summarized by its resilient, flexible and durable character, the risks it faces under environmental threats and its common prevalence around the world. It shows advantages in terms of up-to-date construction materials and techniques, higher load-bearing and spanning capacity, preventing excess building waste by reuse and ease of international cooperation. However, it is also a challenging component due to its complicated legal status as heritage and differences in regulations according to different governments. It can be harder to identify and assess values compared to the vernacular built environment since their authenticity and significance highly depend on their environments and context. Preventative measures regarding the 20th century heritage can also have much more complex technical difficulties since modern materials and techniques can be challenging to implement and costly.

CHOOSING 20TH-CENTURY BUILT HERITAGE FOR ADAPTATION

Flexible and Durable Structures

Under the Threat of Climate Risks with Less Awareness

Common European (and World) Heritage

PROS:	CONS:
- Up-to-date construction materials and techniques - Higher load bearing capacity with larger spaces/spans - Preventing excess building waste - Ease of international cooperation	- Complicated legal status and differences in regulations - Harder to identify and assess values - Higher costs and technical difficulties

Figure 12.1. *The reasons to choose 20th century built heritage for adaptation*

12.2.3. *Methodology of interventions*

After the justification of modern heritage as a suitable asset for adaptation, the process depends on interventions that usually alter the building in the name of adaptive reuse. The differentiation between transforming a building, which has always been a phenomenon, and adaptive reuse is the intention to conserve a heritage site. Therefore, what is crucial is the attitudes in interventions consisting of typological, architectural and technical strategies (Plevoets and Van Cleempoel 2011).

On the topic of adaptive reuse, there are different categorizations of intervention attitudes in the literature. One example that explains these attitudes mentions contemporary adaptive reuse approaches. To briefly summarize, these approaches include the "typological approach", which depends on matching the appropriate function with the proper typology, the "architectural approach", which focuses on the form-to-form relations, the "technological approach", which is when the technical specifications of the building are improved for better performance, the "programmatic approach", which starts by selecting a program and then searching for a historic building to accommodate it, and the "interior approach", which gives value to the immaterial aspects in a romanticist way toward building adaptation (Plevoets and Van Cleempoel 2019). Modern heritage, too, can adapt to climate change with any of these attitudes according to the case requirements.

The criteria for the type of attitude in interventions depend on the values and the condition of the 20th century built heritage. In terms of alteration, there can be additions, removals or renewals in the structural or spatial configuration. However, these alterations should not damage the significance of the heritage, which is the accumulation of its existing values. They should also not affect the current condition negatively. The values can be grouped into physical and contextual aspects. The physical values are related to the form of pure geometry, which is usually determined by the function of the building, and the material of modern technology in construction. Meanwhile, the contextual values depend on the social aspects of the local community and the economic aspects of the function and program of the heritage asset.

On the other hand, the existing condition should be preserved or positively enhanced. In order to make interventions, the authenticity and integrity of the structure or place need to be taken into consideration with utmost care. The architectural and symbolic characteristics of the building define the authenticity that gives it uniqueness. Thus, it is essential to evaluate and repair the changes of human-made or natural causes, which damage physical or contextual values. The integrity of the structure in terms of mass and facade configuration is also essential for its longevity and sustainability during the development of adaptation. This does not mean that there should not be any alterations to the mass and facade, but it emphasizes the importance of a comprehensive method that prioritizes its significance, which is why the case is a heritage asset in the first place. Therefore, the mass and façade alterations should strive to strengthen the integrity that has been weakened or disturbed.

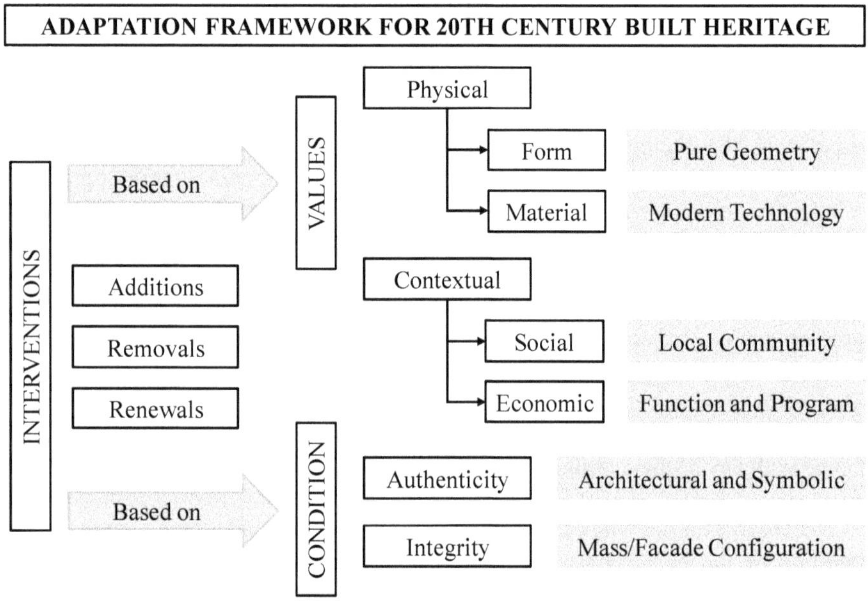

Figure 12.2. *Diagram for the methodology of interventions based on values and condition*

12.2.4. *Development of the framework*

The advantages of 20th century built heritage for climate change adaptation and methods of interventions with regard to certain aspects were prepared to develop a framework, as it was the aim of the study. The adaptation procedure from this framework should increase resilience for the common European heritage and help uplift communities, raise awareness on conservation issues and support sustainable development.

Therefore, the first approach was to identify and evaluate the existing regulations on cultural heritage conservation planning and climate change adaptation. Many regulations were studied, and two of them were determined to be the most suitable for developing a comprehensive general guideline (BSI 2017; ISO 2020). Comparative studies were carried out to determine conflicts or complementary analyses in the planning procedure. Then, based on a consensus, a similar but new framework was established to contribute to the literature on climate change adaptation.

The framework for the adaptation of 20th century built heritage for climate change consists of several actions. These are the identification and documentation of cultural heritage and climate change risks, consideration of the legal context and coordination of the duties of relevant actors, identification and assessment of potential measures with analysis of good examples, development of methodologies for effective policies and management of the schedule, evaluation of the plans and creation of decision-making through engagement, and lastly, a series of actions to complete the process, which includes implementation, monitoring, reporting, sharing of the results and education of future experts.

The first step is identifying and documenting cultural heritage and climate change risks to determine what is being preserved from and adapted to which threats. Documentation is one of the most crucial actions for conservation. Therefore, an inventory of 20th century cultural heritage and the related potential hazards of climate change is essential for obtaining knowledge, preserving information and developing adaptation strategies. The other actions include, but are not limited to assessments of value, vulnerability, risk and impact for further understanding and extending potential strategies for adaptation.

Considering the legal context and coordinating the duties of relevant actors comes as the second step since the documentation should be supported by engaging in dialogue with the stakeholders and decision-makers. The establishment of the conservation project takes place through the identification, investigation and diagnosis, which enables the evaluation and selection of the appropriate conservation options. These options are expected to be compatible with the existing legal context, which consists of laws, regulations, declarations and many more supervisory documents. However, since the policy framework on 20th century architecture is still not definite for some countries like Türkiye, which excludes 20th century buildings in heritage definitions (FAO 1983), it becomes a harder challenge to manage this procedure.

The adaptation needs to be convincing and long lasting. Therefore, the integration of sustainability and international cooperation is expected from the following stages for constant liability and supervision. In that case, identifying and assessing potential measures with the help of analyzing good examples is crucial for keeping up with the latest scientific progress. These measurements should be updated constantly according to the latest findings, and the examples should be checked periodically to assess and document various consequences of the efforts in adaptation.

With the knowledge obtained from the potential measures, developing methodologies for effective policies can be realized through a pre-determined schedule and timetable. As the process continues, evaluation of the ongoing plans helps to provide decision-making through the engagement of several involved parties and stakeholders, like the inhabitants. The management of developing methodologies and decision-making is the final preparation for implementing actions for adaptation.

From there onward, the implementation should be consistently monitored and updated according to the needs of the case area. The entire progress should be documented, and reports should be prepared to explain every action in great detail with pros and cons. It is most beneficial to share the results of these reports with the other cases of 20th century built heritage adaptation to climate change, since the issue needs further discussions based on the results. Moreover, the findings can be the foundation for the education of inhabitants, heritage experts, site personnel, government officials and other stakeholders.

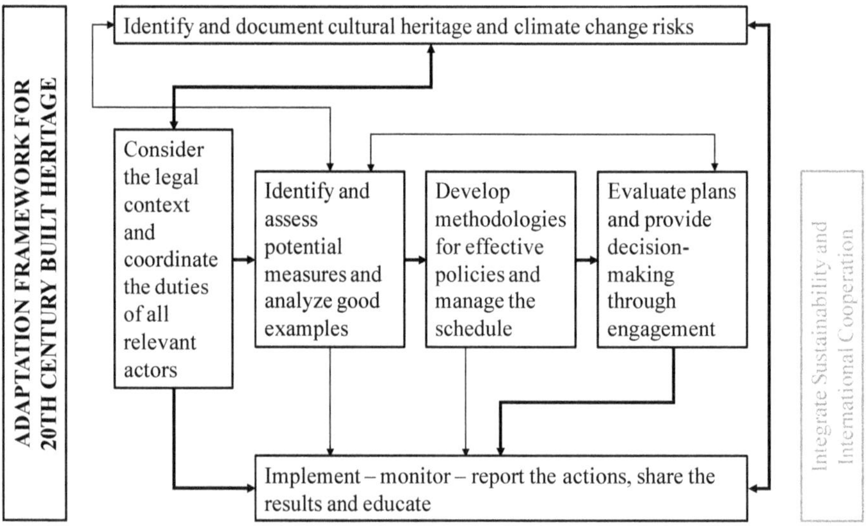

Figure 12.3. *Diagram for the framework of the adaptation process*

Although the framework is set to work in a pre-determined single direction with a particular order, it is also open to different interpretations. Depending on the requirements of the heritage, risks, measurements and consequences, it can be reversed or skipped at various stages, thus offering flexibility and manipulation.

The framework shows a process for a comprehensive adaptation approach, but also highlights the need for dedicated equipment, trained personnel and infrastructure. Thus, it is a very demanding mission to accomplish. However, these are not the only challenges in the face of climate change adaptation. There are many political barriers in the face of bureaucratic processes as well. Since heritage sites are usually not the priority in the policy development agendas of the governing bodies, they are, at the very best, only the secondary focus after people. One of the reasons for this is the issue of budgets and costs. Frequent participation of inhabitants is also ideal but not always a realistic demand (Daly et al. 2022). Therefore, the conservation and adaptation of 20th century built heritage is a privileged topic with far-reaching problems that a strong union like Europe can initiate in the near future.

12.3. Expected results and discussion

12.3.1. *Evaluation of the process*

Heritage is a driver for sustainable development and local communities (ICOMOS 2019). This means that the mere reuse of existing building stock would not be enough to achieve sustainability. There needs to be a regeneration of life in terms of economic, social and cultural development (Yoney et al. 2016). That is why the 20th century built heritage can play a crucial role in adaptation, as it offers a contemporary foundation for today's modern people.

In the discussions of mitigation, cultural heritage can face the barriers of economic factors, lack of regulation, limiting heritage values, inadequate certifications, inefficient use of energy, lack of knowledge, loss of traditional skills, incompatible solutions and the difficulty in identifying shareable solutions (Sesana et al. 2019). When it comes to 20th century heritage, these concerns will also need to be considered from a different perspective. Although it would prove advantageous in terms of obtaining knowledge and achieving compatibility, it could still be a costly and delicate procedure to adapt these structures.

Regardless, the expected results for an adaptation would be to produce solutions for the 20th century heritage to stand against and not contribute to the hazards of climate change. The most prominent benefit would be to generate less building waste of concrete and steel by adapting or reusing them. It would enable the local population to continue to use that space on a building or district scale. Therefore, the further spread of the human population and consumption of building materials and nonrenewable energy resources could be reduced instead of a scenario where people leave their inhabitable neighborhoods. Moreover, new job opportunities for the locals according to the new function of a reused heritage could be achieved. The use of modern materials and techniques would not only increase comfort conditions, but also reduces waste, helps sustainability and boosts resilience against climate change.

Implementing new additions during the adaptation process can make the structure desirable for occupation. Thus, it can revive the interest of the public or people's affection for modern heritage. Their impact on the environment would also decrease if the structures remained in use sustainably to avoid their demolishment or abandonment. This is particularly crucial for most places, since 20th century architecture may not consistently be recognized as a significant cultural heritage due to its relatively young age and lack of traditional techniques.

On the other hand, some studies take an adaptation approach that aims to focus not only on remembering heritage values through permanent adaptation, but also on discovering heritage values that move forward through autonomous adaptation. In that case, there is also the possibility of not adapting a heritage site and celebrating the memory of disturbance caused by climate change (Seekamp and Jo 2020). Although these approaches can be recognized for enriching the discussions in the literature, 20th century built heritage could counterattack the development of resilience if left unattended. Thus, their adaptation may even need to be given priority compared to the traditional heritage.

Nevertheless, there are some aspects in which the 20th century built heritage faces the same issues as other cultural heritage sites. Adaptation plans need to be constantly updated, and they are difficult to implement. Their bureaucratic processes can be slow, and conflicts can frequently occur due to loose definitions in official documents. Concerns about data reliability and the controversial nature of scientific data on climate change are also common concerns. Additionally, since the subject is relatively newer in heritage discussions, there is a lack of information necessary to make fully informed decisions. Therefore, the need for more specialized skills and resources, and different levels of interest and motivation on the subject are apparent challenges in the face of adaptation (Phillips 2014).

12.3.2. *Integration of the outcomes into heritage education*

Contemporary adaptation and mitigation practices are mainly influenced by the natural sciences. Although this involvement is necessary for identifying and understanding the problems created by climate change, it may remain insufficient in the face of knowledge given by cultural heritage to help find cohesive solutions. Therefore, interdisciplinary studies are crucial for developing adaptation, especially for increasing social and cultural resilience (ICOMOS 2019). Mutual learning and coordination form the basis of this interdisciplinary and transnational network. Engaging the younger generation in future capacity building through heritage education becomes a much-needed strategy (European Commission 2019).

In this context, cultural heritage education is part of the compulsory activities undertaken by several states. It is based on different legislations involving various stakeholders and networks. Some organizations, such as UNESCO, have successfully created uniform and evolving education measures that target various stakeholders and focus on multiple levels of awareness (Jagielska-Burduk et al. 2021). In this case, a branch of education focusing on the 20th century built heritage and its adaptation to climate change can be an important idea. Primarily focusing on its potential to unite everyone from different places is essential, since young people may perceive cultural heritage in a limiting, Eurocentric approach (Röll and Meyer 2020).

Several tools are used for innovative communication, learning and teaching methods. They can contribute to developing cultural heritage education through many possibilities, including network creation regardless of geographical limits and using advanced technologies for a unique learning and teaching experience (Ott and Pozzi 2011). Thus, the data on possible outcomes can be shared and exchanged effortlessly. From this perspective, heritage education for adaptation seems to be the primary approach for capacity building regarding training experts and expanding expertise.

12.4. Conclusion and further research topics

When discussing a common European heritage, modern architecture of reinforced concrete and steel cannot be overlooked. Features that define 20th-century structures were born as a reaction to the traditional and promised better living standards with bold interventions. In line with this view, these structures should now be responsible for reducing environmental problems and increasing resilience.

Heritage can contribute to climate debates beyond preserving historic sites, environments and intangible heritage. Recognizing heritage as a study or tool for overcoming global challenges and institutional barriers can unlock broader values and provide critical steps for action, particularly for adaptation (Casey and Becker 2019; Morel and Ammerveld 2021).

Therefore, these heritage assets should be open to bold interventions that will not harm their significance and values. Integrating new technologies, solutions in spatial configuration and architectural elements, and adaptative reuse can be studied to determine appropriate strategies. Since 20th century modern architecture has been constructed worldwide as a common approach in the built environment, it can be the foundation for research of international collaboration for compatibility in the adaptation process, while providing conservation against climate risks.

This study justifies the use of 20th century heritage in adaptation and provides criteria for intervention strategies. Then, these findings are utilized to offer a framework for the adaptation of 20th century built heritage to climate change. However, some aspects still require further research, like the question of how the risk of perpetuating the same flaws of modern architecture could be avoided.

Some study findings also suggest that the adaptation of cultural heritage to climate change is predominantly hindered by institutional and technical barriers, with additional emphasis on sociocultural and financial obstacles. Climate adaptation actions are constrained by institutional barriers such as limited frameworks, standards, technical guidelines, lack of established adaptation prioritization processes, weak cooperation between multi-level actors and competing priorities across sectors (Fatorić and Biesbroek 2020). Unfortunately, in events like these, 20th century heritage would not be prioritized; therefore, more research focusing on institutional barriers and limitations is crucial for further discussions.

Nevertheless, since the existing conditions of the heritage sites would be improved through adaptation, it offers to elevate the situation even if it may not answer all current problems. To work on extensive research, cultural heritage education, training and capacity-building initiatives should be prepared within a broad network.

12.5. References

Bosher, L., Kim, D., Okubo, T., Chmutina, K., Jigyasu, R. (2020). Dealing with multiple hazards and threats on cultural heritage sites: An assessment of 80 case studies. *Disaster Prevention and Management: An International Journal*, 29(1), 109–128.

Bozdoğan, S. (2002). *Modernizm ve ulusun inşası: erken cumhuriyet Türkiye'sinde mimari kültür.* Metis Yayınları, Istanbul.

Brandt, M.T. and Rouillard, C. (2020). Climate chaos and heritage-conservation values: The urgency for action. *APT Bulletin: The Journal of Preservation Technology*, 51(1), 37–48.

BSI (2017). BS EN 16853:2017 standard. Conservation of cultural heritage – Conservation process – Decision making, planning, and implementation [Online]. Available at: https://www.en-standard.eu/bs-en-16853-2017-conservation-of-cultural-heritage-conservation-process-decision-making-planning-and-implementation/.

Casey, A. and Becker, A. (2019). Institutional and conceptual barriers to climate change adaptation for coastal cultural heritage. *Coastal Management*, 47(2), 169–188.

Daly, P., Feener, R.M., Ishikawa, N., Mujah, I., Irawani, M., Hegyi, A., Baranyai, K., Majewski, J., Horton, B. (2022). Challenges of managing maritime cultural heritage in Asia in the face of climate change. *Climate*, 10(6), 79.

European Commission, Directorate-General for Education, Youth, Sport and Culture (2019). European framework for action on cultural heritage. Publications Office, Luxembourg [Online]. Available at: https://data.europa.eu/doi/10.2766/949707.

FAO (1983). Turkish Law No. 2863 on the Protection of Cultural and Natural Assets [Online]. Available at: https://www.mevzuat.gov.tr/mevzuatmetin/1.5.2863.pdf [Accessed 25 December 2022].

Fatorić, S. and Biesbroek, R. (2020). Adapting cultural heritage to climate change impacts in the Netherlands: Barriers, interdependencies, and strategies for overcoming them. *Climatic Change*, 162(2), 301–320.

Fatorić, S. and Seekamp, E. (2017a). Are cultural heritage and resources threatened by climate change? A systematic literature review. *Climatic Change*, 142(1–2), 227–254.

Fatorić, S. and Seekamp, E. (2017b). Securing the future of cultural heritage by identifying barriers to and strategizing solutions for preservation under changing climate conditions. *Sustainability (Switzerland)*, 9(11), 2143.

García-Esparza, J.A. (2019). Engaging architectural heritage in climate action. *VITRUVIO – International Journal of Architectural Technology and Sustainability*, 4(2), VII–VIII.

Hassan, M. and Xie, H. (eds) (2020). Climate change and conservation of coastal built heritage. In *Climate Change and Conservation of Coastal Built Heritage*. Springer, Singapore.

Hedieh Arfa, F., Zijlstra, H., Lubelli, B., Quist, W. (2022). Adaptive reuse of heritage buildings: From a literature review to a model of practice. *The Historic Environment: Policy & Practice*, 13(2), 148–170.

ICOMOS (2019). The future of our pasts: Engaging cultural heritage in climate action. Report, ICOMOS, Paris.

ICOMOS (2021). Triennial Scientific Plan 2021-24 – Cultural Heritage and Climate Action. Report, ICOMOS, Paris.

ISO (2020). ISO/TS 14092 standard. Adaptation to climate change – Requirements and guidance on adaptation planning for local governments and communities [Online]. Available at: https://www.iso.org/standard/68509.html.

Jagielska-Burduk A., Pszczyński M., Stec P. (2021). Cultural heritage education in UNESCO cultural conventions. *Sustainability*, 13(6), 3548.

Le Corbusier (2008). *Toward an Architecture*. Frances Lincoln, London.

Morel, H. and oud Ammerveld, J. (2021). From climate crisis to climate action: Exploring the entanglement of changing heritage in the Anthropocene. *Historic Environment: Policy and Practice*, 12(3–4), 271–291.

Orr, S.A., Richards, J., Fatorić, S. (2021). Climate change and cultural heritage: A systematic literature review (2016–2020). *Historic Environment: Policy and Practice*, 12(3–4), 434–477.

Ott, M. and Pozzi, F. (2011). Towards a new era for cultural heritage education: Discussing the role of ICT. *Computers in Human Behavior*, 27(4), 1365–1371.

Phillips, H. (2014). Adaptation to climate change at UK World Heritage Sites: Progress and challenges. *Historic Environment: Policy and Practice*, 5(3), 288–299.

Plevoets, B. and Van Cleempoel, K. (2011). Adaptive reuse as a strategy towards conservation of cultural heritage: A literature review. In *Wit Transactions on the Built Environment*, vol. 118, Brebbia, C.A. and Binda, L. (eds). WIT Press, Southampton.

Plevoets, B. and Van Cleempoel, K. (2019). *Adaptive Reuse of the Built Heritage: Concepts and Cases of an Emerging Discipline*. Routledge, London.

Röll, V. and Meyer, C. (2020). Young people's perceptions of world cultural heritage: Suggestions for a critical and reflexive World Heritage education. *Sustainability*, 12(20), 8640.

Romão, X. and Bertolin, C. (2022). Risk protection for cultural heritage and historic centres: Current knowledge and further research needs. *International Journal of Disaster Risk Reduction*, 67, 102652.

Seekamp, E. and Jo, E. (2020). Resilience and transformation of heritage sites to accommodate for loss and learning in a changing climate. *Climatic Change*, 162(1), 41–55.

Sesana, E., Gagnon, A.S., Bertolin, C., Hughes, J. (2018). Adapting cultural heritage to climate change risks: Perspectives of cultural heritage experts in Europe. *Geosciences (Switzerland)*, 8(8), 305–328.

Sesana, E., Bertolin, C., Gagnon, A.S., Hughes, J.J. (2019). Mitigating climate change in the cultural built heritage sector. *Climate*, 7(7), 90.

Sesana, E., Gagnon, A.S., Ciantelli, C., Cassar, J.A., Hughes, J.J. (2021). Climate change impacts on cultural heritage: A literature review. *Wiley Interdisciplinary Reviews: Climate Change*, 12(4), 12:e710.

United Nations (2022). What is climate change? *Climate Action* [Online]. Available at: https://www.un.org/en/climatechange/what-is-climate-change [Accessed 25 December 2022].

Yoney, N.B., Asiliskender, B., Ozer, A. (2016). Adaptive reuse of industrial heritage: Resilience or irreparable loss? *DOCOMOMO 14th International Conference: Adaptive Reuse, the Modern Movement Towards the Future*, Lisbon.

13

A Transformative Approach to Cultural Heritage Education for Climate Action

This chapter discusses entry points for including cultural heritage in different planning approaches for climate action and the role of public higher education in transforming practices toward sustainability. It reflects on implicit cognitive biases – registered at the individual level and those embedded in educational curricula – and their impacts on designing systemic planning strategies for societal resilience. This chapter presents two tools for *demystifying complexity* and *connecting the small to the big picture* used in education for sustainable development (ESD) at Leiden University. Their applicability is explored to help future practitioners critically reflect on how "unsustainable" patterns manifest daily in current social systems and reduce the psychological distance and the feeling of hopelessness toward climate action that prevents taking meaningful action. Conclusions discuss how cultural heritage theory and practice can contribute to practicing reflexivity and deliberation, much needed in planning for sustainable climate actions, and the significant role of (formal) knowledge systems in transforming societies toward sustainability.

13.1. Introduction

It is widely recognized that wicked problems like rapid environmental and climate change have interdependent root causes (Sharma 2017; Leichenko and O'Brien 2019; Pascual et al. 2022, to mention a few). However, the limited and fragmented results the world has witnessed in addressing climate mitigation and adaptation shows the need to consider how social and environmental issues are

---

Chapter written by Paloma GUZMAN and Jelle BUIJS.

inextricably linked carefully. The lack of profound action has also raised criticism of current educational systems prioritizing techno-managerial quick fixes and challenged them to instead focus on stewarding societal and technological change in a way that avoids catastrophic impacts. Although formal educational systems have brought major advances in human life and our understanding of contemporary issues, these also perpetuate mindsets and institutional frameworks that shape unsustainable patterns that threaten humanity and the planet today (Bentz and O'Brien 2019).

The seriousness of current socioenvironmental crises has led different scientific fields to recognize that a transformation of science, society and self is inevitable. This chapter aligns with optimistic propositions of how societal transformations can be navigated in educational curricula by expanding our understanding of current climate challenges and the possible impacts of technological solutions. This requires that current knowledge systems enable future practitioners to challenge held assumptions underpinning unsustainable patterns by developing the reflective and deliberate spaces needed to source creative responses to climate challenges (O'Brien 2018; Fazey et al. 2020).

The "transformative turn" has also reached the spatial planning field. It is increasingly advocating for strengthening the connections between the social and natural dimensions. Such calls aim to increase the relationship between societies' cultural and environmental features and the sociocultural values attributed to places to successfully guide climate-resilient development pathways (Horlings 2016; Grenni et al. 2020). Therefore, cultural heritage, both tangible (built environment and objects) and intangible (traditions, practices and ways of knowing), can play a key role in transforming the theory and practice underpinning spatial interventions. Cultural heritage is a social construct that provides communities with identity and a sense of belonging linked to development's environmental, social and economic dimensions (Guzman et al. 2017). Theoretical expansions in cultural heritage definitions can provide insights into different value systems attributed by diverse groups to strategic elements in the natural or built environments (Guzman and Daily 2021).

Consolidating cultural heritage management as a government sector increasingly requires integrating adaptation and environmental strategies (Fatorić and Seekamp 2017; Orr et al. 2021; see Ishizawa et al. 2023 in this volume). At the same time, sociocultural values manifest in spatial attributes in the built and natural landscapes are becoming the backbone for sustainable individual and collective actions in favor of planetary well-being (Bandarin and van Oers 2012; Hakkarainen et al. 2022). Furthermore, the evolving conceptualization of cultural heritage provides a historical perspective on how societies manage local resources. Thus, educating about cultural heritage implications in policy and planning can be a critical leverage point for interdisciplinary and integral climate actions. This chapter explores the possibilities of creating a transformative curriculum in current architecture and planning education by integrating reflexive tools for understanding cultural heritage and its planning nexus in sustainable climate actions. We address the questions: How can architecture education source insights from cultural heritage theory and practice for solving societal challenges related to climate action? How can reflexive and deliberative tools help future cultural heritage practitioners to see the unsustainable patterns underpinning complex societal challenges?

The transformative approach discussed here moves away from technical "fixes" that dominate most educational curricula today. Such practices tend to externalize environmental problems from human agency, that is, the capacity of individuals to recognize their take on the issues but also become the solution. Instead, we discuss how two transformative tools can broaden and deepen cultural heritage education. The tools called *learning full spectrum* and *reflective practice* are currently explored in the Sustainability and Health course, an elective for second-year students at the Honours College track in Science, Society and Self at Leiden University, in the Netherlands. The course takes climate change as an entry point to study the intimate relationship between individual, collective and planetary well-being. This approach enables students to develop mindsets that recognize and value different beliefs, worldviews and paradigms (including their own) on global environmental change and their sociocultural causes. Thus, these tools enhance the agency of future practitioners by enabling them to transform science, society and the self.

13.2. Cultural heritage perspectives for integrative climate planning

Studying the built environment through architecture and spatial planning requires knowledge of pre-existing conditions. It also entails the assimilation of underlying assumptions from institutional and political structures, ideas, uses and values of the built environments, pollution and other forces of the past, including historic structures and practices. Integrating technological innovations in architecture and dealing with different actors' interests in spatial planning has allowed diverse scientific disciplines to cross conventional professional and disciplinary boundaries flexibly. Within these fields of knowledge, cultural heritage management has rapidly developed its own theoretical body expanding cultural heritage definitions and conservation paradigms. As a result, sociocultural knowledge has broadened over time, forging more direct links between cultural and natural resources and planning in terms of sustainable development policy and processes (Veldpaus et al. 2021).

The study of cultural heritage conservation in Western architecture education was motivated by a desire to offset the modernist movement characterized by standardized planning and design proposals through appraising local materials, functions and uses (for a historical overview, see Veldpaus et al. 2013). This conservation paradigm also focused on identifying and documenting historical monuments, which helped forge national imaginaries and identities in the post-war period. Therefore, cultural heritage experts became part of the planning sector that established policies and laws to prevent the loss of historical objects, monuments and areas (Janssen et al. 2017). Related conservation endeavors focus on cultural heritage isolation from spatial dynamics. However, this approach has been criticized for its strong top-down approach and elitist expert-based orientation and for failing to recognize the increasing sociocultural diversity of contemporary societies (Smith 2006; Waterton et al. 2006). As such, it has also contributed to the marginalization of minorities and prioritizing the conservation of materiality over social values and functions.

During the late 1980s and early 1990s, the management practice was expanded when the economic effects of urban regeneration in historic areas led to the positioning of cultural heritage benefits for socio-economic development (Bandarin and van Oers 2012). As a result, fragments from the past started to be assimilated as distinctive elements of larger development strategies (e.g. urbanization, reuse and rehabilitation), contributing to a place's quality. In addition, this approach stimulated collaborations between the public and private sectors. However, such partnerships have proven to respond to market competition, which benefits disregard the disarticulation of local inequality and unjust development patterns (Guzman et al. 2017).

In recent years, a third expansion of cultural heritage understanding is now considering the diversity of stories and meanings associated with artifacts, buildings, districts and nature as a constituent of historical continuity needed for sustainable development (Soini and Dessein 2016). In this emerging paradigm, conserving and the constant re-creation of tangible and intangible heritage guide spatial planning in the broader sense, where interconnections between spatial contexts and sociocultural values and meanings are particularly important (Janssen et al. 2017). Such conceptualization also opens new ways of looking at our relationship with nature and ourselves and others, including more than humans. Understanding such relationships requires attention to the interactions between people inside and outside formal decision-making processes. Cultural heritage becomes a mirror that evidences how different people and groups identify with and attach value to their built and natural environment. It also evidences unsustainable patterns of exclusion, inequity, fragmented collaboration and injustice. However, this requires future practitioners to acknowledge and explore different forms of knowledge to orient decision-making views that embrace local systems' diversity and complexity (Hakkarainen et al. 2022).

The three expanded conceptualizations of cultural heritage have created a planning nexus that can be recognized internationally in management practice. Janssen et al. (2017) synthesized these management practices as heritage as a *sector* (protectionist approach), *factor* (market-based approach) and *vector* (driver of sustainability), which provide current practitioners with a repertoire of tools to contextualize engagement with environmental changes and broader societal values. We argue that the *vector* approach to cultural heritage management can transform practitioners' engagement in current societal and ecological challenges by highlighting sociocultural values and meanings. Emphasis should be given to our societies' contradictory and polarizing values that lead to paralyzing narratives and debates. However, mobilizing shared values through cultural heritage and its management benefits can only be possible when all paradigms underpinning heritage practices question the root causes of inequity, injustice and environmental degradation. Therefore, applying the different paradigms can be seen as a restriction or an opportunity to work with change while realizing the full potential of people and places. Shifting toward the best results for most people depends on the practitioners' reflexive and deliberative capacities to navigate from one approach to another with discernment. Discernment, in turn, requires understanding what is needed to shift in current societal systems that perpetuate the status quo in myriad unsustainable individual and collective practices.

A transformative approach to cultural heritage management implies significant changes in the practices, policies and cultural norms in which societies relate to nature and each other. It entails meaningful collaborations by considering people's motivations to care about and respect nature. Therefore, it demands consideration of the subjectivity of different people's associations with nature and climate actions. To address these challenges, future practitioners will be increasingly challenged to go beyond the dominant scientific positivism that simplifies architects, planners and cultural heritage practitioners' roles as translators and communicators across spatial sectors. Although there is reason to question what heritage experts can do to societal and environmental problems (Veldpaus et al. 2021), we argue that their capacities to rethink causality differently, interpret different value systems and integrate the subjective nature of knowledge should be strengthened to become transformative agents in climate planning.

Enhancing the agency of future practitioners is challenging because governmental departments responsible for managing architectural heritage focus on the technicalities of conservation and maintenance of cultural heritage materiality as their main area of expertise rather than exploring their practice as interconnected with broader climate efforts and goals. The emphasis on material heritage operates with a reactive and protectionist perspective which implies positivist-linear thinking that portrays environmental phenomena (such as climate change) as a threat to conservation activities. Even when reaching more interdisciplinary and collaborative management practices (*factor* approach), the decision-making process remains top-down and expert oriented (e.g. architects, historians and archaeologists). It also increases limiting the role of practitioners to technical advisors on what is to be preserved while avoiding change, and disempowers cultural heritage managers as change agents. Shifting this situation requires more work to integrate climate planning and heritage conservation objectives and to recognize practitioners' potential as strategic coordinators and facilitators.

Moreover, difficulties in operationalizing integrative approaches for climate planning are rooted in sectorized views of the nature of problems. These have led to poor communication among sectors, a need for more awareness of the negative impacts of strategies across the sector and conflicting interests among actors interacting in the spatial context (Guzman and Daily 2021). These recurrent issues make it difficult to implement integrative and coherent solutions that support place-based solutions to global climate issues by considering sociocultural elements to enable cooperation among different worldviews. Therefore, there is an increased recognition of the impacts of other solutions from different perspectives as complementary to enact more creative solutions that effectively enable the urgent societal transformation needed.

13.3. A case study of transformative approaches in higher education

Commonly, higher educational curricula covering climate change issues prioritize a "solid understanding" of science and technology to gain the ability to grasp complex concepts about the climate, climate protection measures and support for climate-resilient development. The applicability of formal knowledge systems (universities and research) can be seen in defining tools and technologies that can support or impede such processes within policy and planning. Often, transformation is explored on practical dimensions of change, that is, it must include all stakeholders equally: if every governmental and public sector person is willing to make consumer behavior and lifestyle changes based on the principles of efficiency and human interest over nature (Fazey et al. 2020).

Conventional educational curricula fall short in driving societal transformations because these derive from scientific silos leading to knowledge separation rather than integration. This outdated, outmoded academia cannot transcend the inertia, failing to break through the current societal crisis. Therefore, creating a transformative integrative curriculum involves recognizing the importance and influence of worldviews, values and beliefs framing climate change problems and solutions (Leichenko and O'Brien 2019). Transformational power is created when collectively we synergistically weave technological results; systems shift as well as the values we all deeply care about (including those universally human, such as equity and justice) (Sharma 2017). This requires that current and future education systems deepen spaces for the individual to think personally and collectively about who we are and how we give meaning to our agency, considering urgent sustainable pathways.

Although (formal) knowledge systems have enabled major societal breakthroughs, such as the invention of the Covid-19 vaccine, we need to be doing better regarding safeguarding the life-support systems of people and the planet. Current knowledge systems, including universities, need to generate new patterns of thinking and acting that are urgently required for life on Earth (Fazey et al. 2020). The neo-liberalized academy has become so unsustainable that it no longer provides the time and space to take a step back and reflect on the nature of today's socioecological crises (Paasche and Ôsterblom 2019). This becomes clear with the common tendency in academia to look for innovative solutions for societal problems. Although there is nothing wrong with looking for solutions, it does become problematic if the problems being addressed remain poorly understood.

A limited understanding of problems leads to partial responses. This often results in technical fixes that fail to address systems and cultural causes (Sharma 2017). A case in point is ESD, which often focuses on the technical side of climate change while neglecting its deeply political and personal dimensions. We thus argue in favor of problem-based rather than solution-oriented education. Although derived from the best intentions, much research and education at universities still operate within the growth-oriented mechanistic paradigm. Given the challenges ahead, important shifts in knowledge systems include from self-referential and uncritical to reflexive and responsible knowledge; from fragmented and disconnected to interconnected and interrelated knowledge; from a focus on the global to a focus on both the global and the local; and finally, from education systems that merely tweak the status quo to education systems that cultivate transformative thinking and acting for everyday lives (Fazey et al. 2020).

Transformative pedagogies that consciously and strategically attempt to shift from existing to envisioned knowledge systems are emerging. This is the case with the Sustainability and Health course at Leiden University, in the Netherlands. Students taking the course are encouraged to think critically and holistically about what climate change can tell us about the *relationships* that constitute our turbulent times and to explore integrated ways forward. Moving away from disempowering doom and gloom scenarios, this course is concerned with reflection upon how we matter more than we think and inquiry into how we can tap into our full potential to contribute to the transformations toward sustainability. Two transformative tools aim to significantly change how students provide meaning to science, society and the self: in the context of climate change by learning the *full spectrum* and engaging in *reflective practices*.

The first tool addresses the technical, social and human dimensions of climate action, enabling students to learn about the full spectrum of this complex problem. To cover the breadth and depth of the topic, the course covers material from various disciplines. This makes students understand that their scientific discipline does matter when it comes to understanding issues like climate change but that other disciplines are also needed to obtain a rich and detailed understanding of the topics. It is, therefore, about shifting from a thinking pattern of either/or to one of both/and. More specifically, students realize that their discipline needs to transform at the depth and speed of the climate to address the broader and deeper questions of the day adequately. This interdisciplinary approach has been enriching to students but sometimes also upsetting because they are confronted with the limitations of their disciplinary knowledge and individual capacities while confronted with their biases and assumptions, which can also be contradictory. Rather than seeing this as a negative outcome, we agree with Biesta (2021) that education is inherently disruptive. The road to humbleness comes with painful yet meaningful learning experiences.

Several reflective exercises are designed to help students to connect the big to the small picture. These are expected to create the basis and habits for reflective practices in future experts and practitioners. Throughout the course, images of The Hague (the city where the classes take place) that illustrate diverse climate change themes were discussed in plenary. For instance, a photograph of a highway near the campus was shown when greenhouse gas emissions were covered. The students were asked to identify the elements in the picture that contributed to the emissions of greenhouse gases. The purpose of this is to make abstract, worldly topics accessible and to show that global problems manifest locally. Another reflective exercise in the course consists of answering reflective questions individually and in a group with students from other disciplines before the session. This exercise allows students to engage with the course materials meaningfully prior to the teacher's explanations. By deliberately reflecting on how the themes addressed do not only manifest "out there" but also "over here", that is, in the everyday life of students, climate change becomes more accessible. This resonates with what Biesta (2021) refers to as the spotlight effect, where students learn first about the world and then their position in it (i.e. being spotlighted) so they can find out what the world may be asking of them. Examples of spotlighting, reflective questions are as follows:

– There is evidence all around us that the climate is changing. What is your strongest memory of an embodied experience of climate change?

– In a session, the class discusses how the mechanistic and holistic visions described by Capra and Luisi (2014) manifest empirically in the world. The teacher requests the students provide at least four concrete examples of how these competing visions manifest in students' everyday life.

– Despite years of warnings by scientists, the Covid-19 pandemic shook many of us and drastically changed our lives. In a session, the class discusses statements by Schipper et al. (2020) on how disruptive events like pandemics scratch on people's perception of the world. Students must then answer the questions: How has the pandemic impacted you? Moreover, how do you make sense of and relate to the world (yourself included)?

– In a session, the class discusses how feminist relational ontology depicts humans as inherently vulnerable and how Leichenko and O'Brien (2019) similarly illustrate vulnerability's multidimensional and contextual nature. Students are asked in the follow-up session: How are you vulnerable in your everyday life, and how do you contribute to the vulnerability of others?

Reflective exercises are thus designed to enhance awareness that climate change is not just an external environmental problem; but a dominant perspective that reduces the issue's complexity. In other words, the students develop a *relationship* with the subject matter. This illustrates one of the key arguments in feminism: that the personal and the political cannot be meaningfully separated (Hanisch 2006). By illuminating how the drivers and impacts of climate change are present in their everyday lives, students can come to terms with the fact that they are not only part of the problem but can also be part of the solution, *depending on how they give meaning to their agency*. Students are, in this way, addressed as subjects rather than objects. More specifically, this educational practice positions students as actors in the local-global political arena. This "subjectification" of the new generation is crucial to transcend the inertia and bring about radical transformation toward sustainability (Sharma 2017; Leichenko and O'Brien 2019; Biesta 2021).

13.4. The role of architecture and planning education in fostering transformation towards sustainability?

Transforming current practices implies a deeper approach to knowledge integration and positions the social sciences and humanities as fundamental for understanding environmental change's social and cultural drivers (O'Brien and Leichenko 2019). There is an increased need for explicit contributions from these fields to understand climate change's social and human dimensions (Tschakert et al. 2016). Particularly, broader engagement with the full spectrum of cultural heritage processes and understandings in the context of climatic change opens possibilities for inclusive dialogues. For instance, technical knowledge on adapting cultural materiality does not consider whether outcomes are right for a particular time and the needs of diverse people. The *sector* approach also sees nature and climate change as threats, and thus, perpetuating the dichotomy between culture and nature. Consolidating cultural heritage management as a *factor* for developing local spatial planning has opened opportunities for broader collaborations among government sectors and private actors. This has facilitated impact and vulnerability assessments of heritage resources and their integration into mitigation strategies through life cycle assessments and the reuse of materials and spaces. However, many times, spatial interventions evidence the perpetuation of exclusionary patterns. Thus, the need for integrative, democratic and just climate action is becoming evident in many fronts of policy and planning (Anguelovski et al. 2016; Veldpaus et al. 2021; Pascual et al. 2022). The need to understand the impacts on cultural heritage sites and their vulnerability has left taking climate action a pending task in practice (see Chapter 15 of this book).

Cultural heritage management is increasingly challenged to understand how broader sociocultural values and benefits are perceived by different social groups, including environmental management, local communities and visitors (Fatorić and Egberts 2020). This gap is also acknowledged in other educational and scientific fields (Fazey et al. 2020). Approaching broader societal values and meanings as a *vector* provides the theoretical basis for social science and humanities to have policy implications. To operationalize more holistic cultural heritage management and climate actions, sharing and understanding multiple ways of knowing is necessary. Future practitioners are increasingly challenged to develop and implement methodologies for breaking disciplinary and sectorial barriers.

This chapter has explored transformative tools that enable practitioners to question taken-for-granted theory and practice in everyday life. Such exercises are useful in shifting perspectives on the technical fixes of climate change toward facilitating sustainable and out-of-the-box responses. Reflective spaces that question how we define vulnerability, resilience and risk scenarios with sociocultural perspectives can provide broader entries for creative collaborations across sectors and actors. Although some think that working with inner capacities to solve problems is simplistic or that we all need technological solutions, becoming a sustainable society requires each of us to embody it, that is to practice sustainability with our minds, hearts and hands in our everyday lives (Sharma 2017). Universities and spaces for formal knowledge creation are increasingly challenged to shift their focus on efficiency to a transformative space for (self-)reflection and dialogue focusing on accomplishment and measurable results. Our proposition is not to create "either-or" divisive views on current knowledge systems. Instead, we see using lived realities and universal human values as a meaningful entry point to climate change as complementary and urgently needed in hard and environmental science and classical pedagogies. We believe that opening spaces for involving pluralistic narratives that bridge the psychological distance toward climate action must become a more common practice at higher education levels. In this effort, educators can contribute to reducing the feeling of hopelessness that disempowers taking climate action.

To conclude, we suggest that further emphasis in education programs should be placed on exercises for collectively establishing common actionable goals and objectives based on the shared humanistic values that foster meaningful collaborations. Additionally, the role of history is critical in understanding social systems in terms of their relational causes and effects. History and the evolving processes of concepts shaping our mindsets help students understand social intentions at different times. Historical resources and approaches can also help to understand conflict and contradictory movements as part of human nature. Thus, for formal knowledge systems to become holistic, they need to offer a perspective on conflict and contradiction not as problems to avoid but as a space for creativity and transcendence. For instance, through the many examples, safeguarding tangible and intangible heritage was upscaled from local activism to impact wider political objectives. In this regard, studying cultural heritage and its role in the built and natural environment provides many theoretical entries to explore global justice and pedagogies of hope.

13.5. References

Anguelovski, I., Shi, L., Chu, E., Gallagher, D., Goh, K., Lamb, Z., Reeve, K., Teicher, H. (2016). Equity impacts of urban land use planning for climate adaptation: Critical perspectives from the Global North and South. *Journal of Planning Education and Research*, 36(3), 333–348.

Bandarin, F. and van Oers, R. (2012). *The Historic Urban Landscape: Managing Heritage in an Urban Century*. Wiley-Blackwell, Chichester.

Bentz, J. and O'Brien, K. (2019). ART FOR CHANGE: Transformative learning and youth empowerment in a changing climate. *Elem. Sci. Anth.*, 7, 52.

Biesta, G. (2021). *World-Centered Education: A View for the Present*. Routledge, New York.

Capra, F. and Luisi, P.L. (2014). *The Systems View of Life: A Unifying Vision*. Cambridge University Press.

Fatorić, S. and Egberts, L. (2020). Realizing the potential of cultural heritage to achieve climate change actions in the Netherlands. *Journal of Environmental Management*, 274, 111107.

Fatorić, S. and Seekamp, E. (2017). Are cultural heritage and resources threatened by climate change? A systematic literature review. *Climatic Change*, 142(1–2), 227–254.

Fazey, I., Schäpke, N., Caniglia, G., Hodgson, A., Kendrick, I., Lyon, C., Page, G., Patterson, J., Riedy, C., Strasser, T. et al. (2020). Transforming knowledge systems for life on Earth: Visions of future systems and how to get there. *Energy Research & Social Science*, 70, 101724.

Grenni, S., Soini, K., Horlings, L.G. (2020). The inner dimension of sustainability transformation: How a sense of place and values can support sustainable place-shaping. *Sustainability Science*, 15(2), 411–422.

Guzman, P. and Daly, C. (2021). Cultural heritage in climate planning. The HiCLIP pilot project for understanding the integration of culture into climate action. A report on the Climate Heritage Network WG4 HiCLIP project. Report, ICOMOS.

Guzman, P., Roders, A.R.P., Colenbrander, B.J.F. (2017). Measuring links between cultural heritage management and sustainable urban development: An overview of global monitoring tools. *Cities*, 60, 192–201.

Hakkarainen, V., Soini, K., Dessein, J., Raymond, C.M. (2022). Place-embedded agency: Exploring knowledge–place connections for enabling plurality in governance of social–ecological systems. *People and Nature*, 4, 1141–1158 [Online]. Available at: https://doi.org/10.1002/pan3.10365.

Hanisch, C. (2006). The personal is political. Introduction [Online]. Available at: https://www.carolhanisch.org/CHwritings/PersonalIsPol.pdf.

Horlings, L.G. (2016). Connecting people to place: Sustainable place-shaping practices as transformative power. *Current Opinion in Environmental Sustainability*, 20(May), 32–40.

Leichenko, R. and O'Brien, K. (2019). *Climate and Society: Transforming the Future*. Polity Press, Cambridge.

O'Brien, K. (2018). Is the 1.5°C target possible? Exploring the three spheres of transformation. *Current Opinion in Environmental Sustainability*, 31, 153–160.

Orr, S.A., Richards, J., Fatorić, S. (2021). Climate change and cultural heritage: A systematic literature review (2016–2020). *The Historic Environment: Policy & Practice*, 12(3–4), 434–477.

Paasche, Ø. and Österblom, H. (2019). Unsustainable science. *One Earth*, 1(1), 39–42.

Pascual, U., McElwee, P.D., Diamond, S.E., Ngo, H.T., Bai, X., Cheung, W.W.L., Lim, M., Steiner, N., Agard, J., Donatti, C.I. et al. (2022). Governing for transformative change across the biodiversity–climate–society nexus. *BioScience*, *72*(7), 684–704.

Sharma, M. (2017). *Radical Transformational Leadership: Strategic Action for Change Agents*. North Atlantic Books, Berkeley.

Soini, K. and Dessein, J. (2016). Culture-sustainability relation: Towards a conceptual framework. *Sustainability*, 8(2), 167.

Smith, L. (2006). *Uses of Heritage*. Routledge, London.

Tschakert, P., Tuana, N., Westskog, H., Koelle, B., Afrika, A. (2016). TCHANGE: The role of values and visioning in transformation science. *Current Opinion in Environmental Sustainability*, 20, 21–25.

Veldpaus, L., Pereira Roders, A., Colenbrander, B.J.F. (2013). Urban heritage: Putting the past into the future. *The Historic Environment*, 4(1), 18–33.

Veldpaus, L., Kisić, V., Stegmeijer, E., Janssen, J. (2021). Towards a more just world: An agenda for transformative heritage planning futures. In *A Research Agenda for Heritage Planning: Perspectives from Europe*, Stegmeijer, E. (ed.). Edward Elgar Publishing Ltd., Cheltenham.

Waterton, E., Smith, L., Campbell, G. (2006). The utility of discourse analysis to heritage studies: The Burra Charter and social inclusion. *International Journal of Heritage Studies*, 12(4), 339–355.

14

Recalibrating Seasonal Cultures with Communities

14.1. Seasonal cultures as intangible heritage for climate adaptation

Seasonal cultures and calendars are one of the earliest frameworks communities have used to make sense of the world and time (Roncoli et al. 2008). Through seasonal cultures, people have long tracked and synchronized sets of social and environmental rhythms, patterning community activities over the year (Kwiecien et al. 2021). Indeed, recent interpretations of the Lascaux cave paintings suggest that ice age hunter-gathers used systematic calendars to record ecological phenomena over a year, such as the seasonal behavior of animals[1]. From a cultural standpoint, we can discuss seasonal cultures as formal and informal frameworks tied to human activity, signaling and structuring *individuals' and groups' perceived patterns of yearly rhythms that they segment into periods meaningful for them, and effect practices that maintain or change these patterns* (there is interdisciplinary literature on this topic. See, for example: Adam 1998; Edensor 2006; Birth 2013; Chisholm Hatfield et al., 2018; Krause 2013; Wersan 2017). Orlove (2003) asserts that today, most cultures worldwide divide the year seasonally in some way.

Seasonal cultures remain influential in contemporary societies. These cultures continue to tie daily life to the rhythms of the environment and agriculture. For some indigenous communities (Leonard et al. 2012; Chisholm Hatfield et al. 2018), this link can be explicit, while in urban communities this link may surface less obviously

---

Chapter written by Scott BREMER.

1 Available at: https://www.theguardian.com/science/2023/jan/05/amateur-archaeologist-uncovers-ice-age-writing-system.

(Hitchings 2010), for example, in clothing customs or coinciding school holidays with historic periods of farm labor. But seasons are likewise marked out by social rhythms, like sports, tax years or religious events; they are not only associated with "natural" phenomena. While seasons are uniquely experienced by local populations, they are composed of rhythms at a panarchy of spatial and time scales, from global rhythms like climatic circulations or international markets, down to rhythms specific to a place like local festivals or hydrology. And in this way, seasons can have a general meaning or a very specific and practical meaning for the activities of groups in places: the way hunting season guides hunters or tourist season hoteliers (Orlove 2003; McNeeley and Shulski 2011). Seasonal cultures can be repertoires of symbols, stories, practices and worldviews for strategizing and justifying timely action (Swidler 1986; Sewell 1992).

Some seek to preserve seasonal cultures as heritage to protect them from being undermined by intersecting change processes, from modernization to urbanization, globalization to colonialization, or climate and environmental change (Nas 2002). Communities see their traditional seasonal patterns of thinking, feeling and acting unravelling, and cast around for ways to codify them and make them explicit. As *tangible* heritage, we see calendars and seasonal representations at sites of cultural significance, such as the Lascaux caves. But seasonal cultures are better captured as *intangible* heritage;

> the practices, representations, expressions, knowledge, skills – as well as the instruments, objects, artefacts and cultural spaces associated therewith – that communities, groups and, in some cases, individuals recognize as part of their cultural heritage (UNESCO 2003, Article 2.1).

Seasonal cultures' heritage status has been made concrete through efforts to register them on UNESCO's World Heritage List (Smith 2006). In 2016, UNESCO registered the Traditional Chinese Seasonal Calendar as intangible heritage (Yang 2018), and there have been more recent bids. Project Mausam in India seeks to inscribe the monsoon patterns, cultural routes and maritime landscapes of the Indian Ocean on the UNESCO World Heritage List (Ray 2019). And in 2020, an initiative by outdoor clothing manufacturer Bergans and the World Wildlife Fund sought to "Save the Seasons" by recommending them for the World Heritage List. These examples highlight the politics of recognizing seasonal heritage, rooted in who has the power to elevate their seasonal calendar as deserving of protection.

I argue that seasonal cultures should not be protected against change, but evolve as living cultural repertoires; adjusting to processes like climate change, while providing resources for communities to adapt and transform with such changes. This means recognizing that intangible cultural heritage is "constantly recreated by communities and groups in response to their environment" (UNESCO 2003, Article 2.1). Seasons culturally evolve through public discourse on what is meaningful and practical in a changeable world (Smith 2006; Strauss 2012) and are fundamentally frameworks for adaptation (Kwiecien 2021). They correct routines and practical schemas to fit seasonal cycles, whether adjusting when seeds are sown or the management of a city's stormwater system. Engaging with – learning about, training and playing with – seasonal cultures is one way communities can build capacity for climate action. We need to recalibrate seasonal cultures to organize and coordinate human activity relative to shifting phenological cycles, climatic patterns, societies, technologies, cultures and policies in our communities' institutions (Bremer et al. 2021, 2023).

In this chapter, I present collaborative research that aimed to nurture critical reflection on the seasonal cultures affecting activities and recalibrate these cultures to sustainably adapt to change. I engage with seasons as culture, as a departure from the heritage literature, and look less at processes of education than at interventive research practices. That is, the research did not teach groups about seasonal cultures as adaptative capacity, but rather created the conditions for groups to uncover what seasons concretely mean and do in their contemporary activities, and ways in which they could alter their cultural relation to seasonality. The chapter describes the methods utilized in this research approach.

14.2. Co-producing seasonal cultures within institutions

CALENDARS[2] was a participatory research project studying seasonal cultures in collaboration with groups acting and interacting in social worlds, within *institutionalized* fields of activity from gardening to teaching. The project explored how people are socialized around activities according to unique sets of worldviews, routines, rules and norms, including seasonal schema that set a rhythm to activities and indicate the time for certain tasks (Sewell 1992; Scott 2014). We posited that it is in institutions that seasons become most visible and tangible as repertoires influencing people's behavior, including how they adapt to changing climatic and other rhythms. Rather than seeing seasonal cultures as historical, we were interested in the ways these cultures are contested and changed, and to what extent this builds adaptive capacity. CALENDARS case studies were global and diverse, including beekeepers, botanical gardeners, coffee roasters, artists, climate scientists and schoolteachers and students in Bergen, Norway; grape growers in New England, USA; meteorologists in Ethiopia; amateur writers, conservationists, farmers,

---

2 CALENDARS *(Co-production of Seasonal Representations for Adaptive Institutions)* was a five-year project running from 2019 to 2023, funded by the European Research Council, and conducted by a six-person team of interdisciplinary social scientists, mainly spread over Norway and New Zealand.

hunters, schoolteachers and students and local government on the Coromandel Peninsula in New Zealand.

The CALENDARS study took a unique shape in each case study, but was guided by a common goal of co-producing changes to how groups understood and enacted seasonal cultures. We triggered processes of critical reflection within groups about the complex and intersecting changes – including climatic change – affecting seasonal patterns, and ways in which they adapt their practices accordingly. We created settings for social learning, including appraisal of how fit-for-function their activities are, relative to seasonal rhythms (see Meisch et al. (2022)). The desired impact was for groups to revise their sense of seasonality and devise high-quality cultural repertoires for adapting activities to seasonal variability and change. For example, some Western Norway beekeepers perceive that the early onset of spring temperatures is giving rise to mismatches between the timing of bee foraging, flowering and the legal and practical routines of setting up hives for fruit tree pollination. In response, beekeepers are revising routines for feeding bees, strengthening them for an ever earlier pollination season. In altering patterns of apiculture, beekeepers are co-producing their seasonal cultures for climate adaptation.

This mode of research saw the researcher embedded in institutions, but playing the role of the outsider who draws attention to seasonal rhythms usually taken-for-granted and routinized. In questioning rhythms, researchers brought these rhythms into focus and made them topics of conscious analysis and discussion among groups. People sometimes contested established schemas of activity, noted how they have changed with conditions and put forward alternative ways of thinking and acting seasonally. People came to recognize their agency in altering institutionalized patterns of activity. In this mode, the research was inspired by different strands of work, including "critical ethnography" (Madison 2011), "embedded humanists" (Fisher and Schuurbiers 2013) or notions of "performative" research (Loxley 2007).

Effecting this research meant collaboratively, creatively and iteratively developing approaches to make seasonal cultures visible and changeable. Methods were fitted to each social world, and the ways in which groups interact in those worlds; for example, while the gardeners routinely work together, beekeepers mainly work individually and meet less frequently. Methods had to be sufficiently flexible to admit different ways of knowing and acting seasonally; some seasonal representations could be voiced in interviews, or written on paper templates, while other practices had to be observed, or experienced. Methods had to bridge knowledge and action, with a view to affecting transformative change in institutions; in re-learning the seasonal patterns, what effect did this have on peoples' schemas for action? Methods also had to be determined as credible and legitimate by collaborators.

14.3. Mixed methods for studying and changing seasonal cultures

14.3.1. *Critical ethnography*

One CALENDARS approach was for researchers to embed themselves as ethnographers within institutions, with me embedding myself in an Arboretum, for example. I worked as a volunteer gardener for 50 Fridays over two years, working alongside and interacting closely with the 10 professional gardeners and three scientific staff working there. Over this period, I drew on complementary methods, which mirrored those of other researchers on CALENDARS.

As a *participant observer*, I kept notes on how I experienced seasonal variation through my work in the gardens; from pruning roses in the sun to clearing drains under pouring rain, repotting plants in the greenhouse or felling trees in the snow. In discussing my observations, I asked co-workers if their seasonal experiences were similar, and in this way elicited their seasonal patterns of experience. At the same time, as a trainee gardener, I had a lot to learn, and through the incessant questioning of "why" tasks were completed when they were, I encouraged my co-workers to question routines themselves. Similarly, I recorded the practical patterns of co-workers, and repeated these observations back to them at later moments, asking them if my observations held as typical. This iterative process of observation and discussion was complemented with *semi-structured interviews*, where I asked individuals to tell the story of their year and how they bracketed seasonal periods (if indeed they did). Framed as "narrative interviews", the questioning began from seasons as typified – an imagined "average" year – but then urged interviewees to recall specific examples of seasonal phenomena. This exercise began from the realization that while cultures may have a certain rigidity, we accept that seasons repeat in an altered way, so that no two summers are the same. Interviews also revealed matters – such as values – which were not accessible through observation alone. These interviews too became material for reflexive conversations with co-workers at later points.

I also stimulated reflection through images. Each day I worked at the gardens I took two *photos* from the same points at the same time of the day. This created a photo series of seasonal cycles in the gardens, reflected in the light, phenology, weather, visitor numbers, hydrology and so on. I printed out large time series and asked co-workers to interpret which month photos were taken in, and which cycles or signals they paid attention to when judging the time of the year. I also took photos of the environment and activities I was engaged in each day and asked co-workers to take photos of me at work. In this way, co-workers chose what was interesting to capture in the photo, whether it was the landscape I worked in, or the tools in my hands, for instance. In a related initiative by a CALENDARS researcher working with conservationists in New Zealand, she took a *body-mounted camera* and recorded the two- to four-hour walks that she conducted with her co-workers, re-baiting traplines for predators, capturing her conversation and the movement through the landscape with the interlocutor. These videos formed part of the materials presented back to actors for discussion.

Two other exercises saw co-workers codifying seasonality in *calendars* and in *meteorological measurements*. I provided co-workers with circular calendar templates, marked with 12 Gregorian months like slices of a pie, in a clockwise fashion, as well as the longest and shortest days of the year: the summer and winter solstices. Over some months, co-workers were asked to note down on the calendar: (i) events that mark their year; (ii) tasks at different times of their year; (iii) signs that they used to mark the passing of the year; and finally, reflecting on these marks on their calendar, they were asked to (iv) segment the year into seasonal periods where relevant. This proved to be a challenging task for some, but in placing the calendars alongside each other, some common seasonal reference points became clear, around which co-workers coordinate their diverse seasonal activities. Many institutions we studied either already had a weather station or were supplied with them. This gave them access to a constant stream of meteorological data. In the Arboretum, reading these data was part of the daily routine and gave co-workers pause to consider how they worked by rhythms of temperature, wind and precipitation. They also reflected on how forecasts – to the hour – patterned their daily work.

Finally, central to co-producing the research as legitimate to groups was ensuring they steered the study in directions of interest and use for them. Some groups laid out, from the outset, research outcomes they wanted to steer toward. In the Arboretum, for example, co-workers asked me to maintain a *blog* where I periodically noted down reflections on themes that stood out for me. These blogs were read by co-workers at the arboretum, and this became another channel through which their seasonal experience was reflected to them. I wrote short vignettes about certain activities I did, which co-workers had to situate in time. As another example, the researcher embedded one day every two weeks in a primary school in Bergen was asked by the teacher to give short talks on topics like social science and seasonality, and help take the class on excursions themed to the seasons. At the end of the school year, the researcher took the class to a workshop exercise making contemporary calendars.

14.3.2. *Workshops and walkshops*

CALENDARS researchers organized workshops with a focus on (re-)drawing activities calendars. These were group exercises, where calendars were considered not as pre-set and fixed, but as fluid, contested and malleable. To continue the example of the Arboretum, we designed a workshop that started from the inconsistencies between gardeners' and scientists' calendars – even of individuals engaged in the same activity – before moving on to find what is common; the shared seasonal reference points around which the group coheres and coordinates. We tasked co-workers with designing a contemporary "primstav" – or traditional Norwegian wooden plank calendar – for the Arboretum, in a workshop that functioned as a "design-probe" (see Paschetz et al. (2022)). We used the same blank calendar templates as for the individual exercise, but asked participants to fill out these calendars in groups, recording their debates and the choices they made about which seasonal markers to include or not. This saw co-workers agree on 25 symbols that mark the passing of the year and anchor seasonality – from school holidays to flowering, work tasks to festivals – which they structured to the primstav rectangular format. Co-workers supplied images representing those symbols and found wood from felled trees at the Arboretum from which the calendars will be milled. The complete calendars will be put up at the gardens – in the offices and workshops – and presented at public events.

Another CALENDARS researcher – working with schools in New Zealand – employed "walkshops" as a method. The researcher accompanied primary school students and their teacher on walks over the same stretch of river and forest at different times of the year, and identified seasonal markers on each walk, making environmental rhythms noticeable. The goal was to develop students' habit of noticing, and connection to seasonal patterns, and record and analyze what signals they noticed. With secondary school students, the researcher helped organize a mix of outdoor activities at different times of the year, from camping in the bush in the winter to scuba diving in the summer, with the aim of reflecting on the seasonality of activities; why activities are timed and coordinated as they are. What cycles are activities attached to?

14.3.3. *Art-science collaboration*

Insofar as the CALENDARS team saw seasons as cultural, we argued that they were best made visible and changed in culturally meaningful ways. We connected the research to the creative arts as a medium for rethinking common-sense conceptions of the world, where we saw the arts broadly as inclusive of visual artists, writers and filmmakers, for example. This meant forming mutually influential relationships with artists at the beginning of the project and maintaining those relationships over the long term; the scientific work served as inspiration for artistic representations, while those representations simultaneously served as material for scientific analysis. The intention was to provide participants across our many case studies – and the public – with reflexive experiences (co-produced with the creative arts) from which new patterns of seasonal action arise, at the same time as new conceptions and empirical study of seasonal change. In this way, the project forged close links between science and art, but also across the case studies. In Bergen, for example, the artists worked closely with the Arboretum to organize an art exhibition at the gardens.

In New Zealand, researchers worked most with an amateur writers' group and a filmmaker. The writers' group had already existed for a long time and consisted of Coromandel Peninsula residents who enjoyed writing creatively as a hobby and met to share what they had written, mainly short stories and poetry. Some of their writing features in the local newspaper. Researchers linked up to this group suggested they write about "changing seasonality", and met with them regularly to discuss the theme. This gave rise to a rich store of reflexive creative writing about Coromandel seasonalities, some of which were shared locally. The filmmaker is an award-winning documentary-maker, who shadowed the researchers on some of their walkshops, interviews and other ethnographic work. He documented the research and how people act seasonally, and produced a film-length documentary to be shown at film-festivals and locally. While the film follows the researchers, it is an independent piece of work, telling the filmmaker's own narrative of rethinking seasons with local communities.

In Norway, researchers collaborated with a group of visual artists who produced a virtual reality (VR) simulation of Bergen. When putting on the VR headset, users find themselves in the valley where Bergen is situated, but all vestiges of human habitation are stripped away; the buildings and roads and farms are all replaced by a "pristine" natural landscape. In this simulation, users can manipulate time with certain physical gestures, changing the time of day, the season or the weather, putting themselves in a forest on a cold winter's night, a summer rain shower or the sun of a spring morning. This way of being in all of the seasons at once is a reflexive experience that has proven, at several exhibitions and workshops, to fascinate users and draw out reflexive comments about what marks seasons in nature.

14.4. Affecting adaptation through seasonal cultures

I have shown that seasonal cultures – as intangible heritage – have long served as important repertoires or capacities for adapting to seasonal variability and change. But for them to continue to serve this adaptive function, seasonal cultures must be seen as dynamic frameworks that people can (and do) change according to the shifting rhythmic patterns that communities live by, patterns that morph with changing climates, environments, societies, cultures, technologies and so on. How communities adjust seasonal cultures will determine if they guide communities down adaptive or maladaptive paths.

This chapter showed how a research project can work to "relearn" seasons and recalibrate cultures with groups of people engaged in institutionalized fields of activity. This demands innovation in how research is designed, conducted, analyzed and used in social worlds. The social sciences and humanities have an important role in critically questioning the totalizing and taken-for-granted seasonal categories that are being held in place by the elite – from climate scientists dividing data into temporal periods to groups registering seasons on the World Heritage List. Research can encourage individuals and groups to think about what seasons mean for them, how these meanings are contested and change, and (importantly) how they can affect change themselves. The CALENDARS project was in many ways a pilot, working across a few local-scale cases, but we have already seen evidence of how groups are rethinking and adjusting their seasonal routines because they have taken part in this study.

14.5. Acknowledgments

This chapter was written as part of the CALENDARS project, which received funding from the European Research Council (ERC) under the European Union's Horizon 2020 research and innovation program (Grant agreement 804150).

14.6. References

Adam, B. (1998). *Timescapes of Modernity: The Environment and Invisible Hazards*. Routledge, New York.

Birth, K. (2013). Calendars: Representational homogeneity and heterogeneous time. *Time & Society*, 22(2), 216–236.

Bremer, S., Glavovic, B., Meisch, S., Schneider, P., Wardekker, A. (2021). Beyond rules: How institutional cultures and climate governance interact. *WIRES Climate Change*, 12(6), e739.

Bremer, S., Stiller-Reeve, M., Mamnun, N., Lazrus, H. (2023). Co-producing representations of summer rainfall in Bangladesh. *Regional Environmental Change*, 23(2), 60.

Chisholm Hatfield, S., Marino, E., Whyte, K.P., Dello, K.D., Mote, P.W. (2018). Indian time: Time, seasonality, and culture in traditional ecological knowledge of climate change. *Ecological Processes*, 7(1), 1–11.

Edensor, T. (2006). Reconsidering national temporalities: Institutional times, everyday routines, serial spaces and synchronicities. *European Journal of Social Theory*, 9(4), 525–545.

Fisher, E. and Schuurbiers, D. (2013). Socio-technical integration research: Collaborative inquiry at the midstream of research and development. In *Early Engagement and New Technologies: Opening up the Laboratory*, Doorn, N., Schuurbiers, D., Van de Poel, I., Gorman, M.E. (eds). Springer, Dordrecht.

Hitchings, R. (2010). Seasonal climate change and the indoor city worker. *Transactions of the Institute of British Geographers*, 35(2), 282–298.

Krause, F. (2013). Seasons as rhythms on the Kemi River in Finnish Lapland. *Ethnos*, 78(1), 23–46.

Kwiecien, O., Braun, T., Brunello, C.F., Faulkner, P., Hausmann, N., Helle, G., Breitenbach, S.F. (2021). What we talk about when we talk about seasonality – A transdisciplinary review. *Earth-Science Reviews*, 255, 103843.

Leonard, S., Parsons, M., Olawsky, K., Kofod, F. (2013). The role of culture and traditional knowledge in climate change adaptation: Insights from East Kimberley, Australia. *Global Environmental Change*, 23(3), 623–632.

Loxley, J. (2007). *Performativity*. Routledge, Oxon.

Madison, D.S. (2011). *Critical Ethnography: Method, Ethics, and Performance*. Sage Publications, Thousand Oaks.

McNeeley, S.M. and Shulski, M.D. (2011). Anatomy of a closing window: Vulnerability to changing seasonality in Interior Alaska. *Global Environmental Change*, 21(2), 464–473.

Meisch, S.P., Bremer, S., Young, M.T., Funtowicz, S.O. (2022). Extended peer communities: Appraising the contributions of tacit knowledges in climate change decision-making. *Futures*, 135, 102868.

Nas, P.M. (2002). Masterpieces of oral and intangible culture: Reflections on the UNESCO World Heritage List. *Current Anthropology*, 43(1), 139–148.

Orlove, B. (2003). How people name seasons. In *Weather, Climate, Culture*, Strauss, S. and Orlove, B.S. (eds). Berg, Oxford.

Paschetz, L., Bastian, M., Bowler, R. (2022). Revealing social infrastructures of time. In *Speaking for the Social: A Catalogue of Methods*, Knox, H. and John, G. (eds). Punctum Books, New York.

Ray, H.P. (2019). "Project Mausam". India's transnational initiative: Revisiting UNESCO's World Heritage convention. In *Travelling Pasts: The Politics of Cultural Heritage in the Indian Ocean World*, Schnepel, B. and Sen, T. (eds). Brill, Leiden.

Roncoli, C., Crane, T., Orlove, B. (2008). Fielding climate change: The role of anthropology. In *Anthropology and Climate Change: From Encounters to Actions*, Crate, S. and Nuttall, M. (eds). Left Coast Press, Inc., Walnut Creek.

Scott, W.R. (2013). *Institutions and Organizations: Ideas, Interests, and Identities*. Sage Publications, Thousand Oaks.

Sewell Jr., W.H. (1992). A theory of structure: Duality, agency, and transformation. *American Journal of Sociology*, 98(1), 1–29.

Smith, L. (2006). *Uses of Heritage*. Routledge, Oxon.

Strauss, S. (2012). Are cultures endangered by climate change? Yes, but... *Wiley Interdisciplinary Reviews: Climate Change*, 3(4), 371–377.

Swidler, A. (1986). Culture in action: Symbols and strategies. *American Sociological Review*, 51(2), 273–286.

Wersan, K. (2017). The early melon and the mechanical gardener: Toward an environmental history of timekeeping in the long eighteenth century. *Environmental History*, 22(2), 282–310.

Yang, X. (2018). Chinese traditional perceptions of the calendar year: Implications of Jieqi for contemporary product development and sustainability. *Tourism Management*, 64, 202–217.

15

Integrating Climate Change Concerns in Capacity Building for World Heritage

In 2011, the World Heritage Heritage Committee approved the *World Heritage Capacity Building Strategy* (WHCBS), where the learning approach shifted from training to developing capacities on the emerging needs within the World Heritage system. Replacing the *Global Training Strategy for Cultural and Natural Heritage* (World Heritage Committee 2001), the WHCBS focuses on strengthening competencies and skills of practitioners, institutions and communities, expanding from an approximation to the individual professional to a collective process fostering peer learning and scaling up activities through networks that link heritage with the wider environment. The WHCBS identifies heritage isolation, both in terms of managing sites as islands and working as practitioners disconnected from other sectors, as one of the main problems for the advancement of heritage conservation (UNESCO 2011, p. 2). Thus, the strategy promotes a wider approach to heritage management, as well as the creation of strong joint networks for cultural and natural heritage professionals, bridging the nature-culture divide. The WHCBS maintains that

> (…) successful methodologies and positive lessons learned from each type of heritage can be used to reinforce the protection of the other, while also respecting the specific conservation needs and concerns of each (UNESCO 2011, p. 4).

These ideas served as premises for the creation of World Heritage Leadership (WHL), a capacity building program launched in 2016 and led by the International Centre for the Study of the Preservation and Restoration of Cultural Property (ICCROM) and the International Union for the Conservation of Nature (IUCN) as

---

Chapter written by Maya ISHIZAWA ESCUDERO, Eugene JO and Rohit JIGYASU.

Advisory Bodies to the World Heritage Committee in implementing the United Nations Educational, Scientific and Cultural Organization (UNESCO) 1972 *Convention concerning the Protection of the World Cultural and Natural Heritage* (hereafter the Convention). This program is achieved in collaboration with the International Council on Monuments and Sites (ICOMOS) and the UNESCO World Heritage Centre, funded by the Ministry of Climate and Environment of Norway, as well as other partners[1]. Even though climate change was already recognized as a major factor impacting the protection of World Heritage properties and was already part of the UNESCO World Heritage Centre programs, in particular in relation to natural properties, it was not considered as a capacity building theme in the WHCBS but part of disaster risk management (DRM) aspects. However, since 2011, it has been increasingly recognized as one of the major threats to the outstanding universal value of all types of properties by State Parties to the Convention, World Heritage site managers, heritage experts and Indigenous and local communities reporting on their state of conservation (UNESCO 2021). Thus, climate change is now a fundamental theme for capacity building that the WHL is addressing in its second phase starting in 2023.

In this chapter, the authors will first provide a background to how climate change concerns have been increasingly integrated in the implementation of the World Heritage Convention at policy, research and practice levels. Then, the main shifts in the capacity building approach for World Heritage will be discussed: from a focus on the individual to enabling peer learning, from a focus on sectorial silos to the provision of interdisciplinary and multi-disciplinary learning environments, and from a property centered to a heritage place approach to management and conservation. In this process, we will highlight how integrating climate change concerns has become not only urgent, but also possible due to these paradigm shifts, providing examples of the activities that the WHL has been undertaking so far on this topic. Finally, we will discuss the way forward suggesting areas of work based on lessons learned from previous capacity building experiences.

15.1. Climate change concerns in the World Heritage system

Impacts of climate change on World Heritage properties were first noted by the World Heritage Committee in 2005 when organizations and individuals raised concerns in relation to Sagarmantha National Park (Nepal), Huascaran National Park (Peru), Great Barrier Reef (Australia) and Belize Barrier Reef Reserve System (Belize) (UNESCO 2005), requesting that these natural properties be added to the List in Danger. Since then, several reports (UNESCO 2007a), research (Leissner and

1 Korea and Switzerland.

Kilian 2014; Guzman et al. 2020; Seekamp and Jo 2020), case studies (UNESCO 2007c) and guidances (UNESCO et al. 2010; UNESCO 2014) have been published and activities have been undertaken for raising awareness and understanding better the impacts of climate change on heritage, as well as devising possible solutions. Furthermore, the *Policy on the Impacts of Climate Change on World Heritage* was adopted in 2007 (UNESCO 2007b) and is currently being updated (UNESCO 2021). The 2007 *Policy* already raises attention on the need to develop tools and methodologies to help managers[2] in particular to assess climate change vulnerability and guidance to plan mitigation and adaptation for their World Heritage sites. In the updated policy, this aspect is being reinforced with the shift from a focus on understanding impacts to taking climate action, which includes among other key priorities, knowledge sharing, capacity building and raising awareness (UNESCO 2021). While initially most of research and capacity building had been developed in the context of natural heritage (UNESCO 2011; Heron et al., 2017; UNESCO et al. 2021), ICOMOS assembled a working group on Climate and Heritage that in 2019 published a report outlining the current and potential impacts of climate change on cultural heritage and highlighting that all types of cultural heritage including archaeological sites, historic cities, cultural landscapes and vernacular heritage, and, in particular, coastal heritage and intangible cultural heritage are not only highly vulnerable to climate change but that impacts are already being reported (ICOMOS 2019).

In addition, further work has been undertaken in order to liaise the World Heritage Convention with other relevant conventions and organizations dealing directly with climate change issues, including the United Nations Framework Convention on Climate Change (UNFCC) and the Intergovernmental Panel on Climate Change (IPCC). A joint international meeting between UNESCO, the IPCC and ICOMOS took place in December 2021[3], which further pointed at the need to work with diverse knowledge systems, including Indigenous and local knowledge for addressing climate change issues, and seeing culture and cultural heritage as both a vulnerable asset but also a potential solution to mitigate and adapt to impacts (Orlove et al. 2022; Sheperd et al. 2022; Simpson et al. 2022). The need for further research on impacts and solutions and particularly in methodologies for breaking disciplinary and sectorial barriers, as well as strengthening the policy-science-practice links, was highlighted in the *Global Research and Action Agenda* proposed as an outcome of this meeting (Morel et al. 2022).

---

2 According to the World Heritage Leadership Glossary (forthcoming), managers are the institution(s) or other type(s) of entity(ies) and group(s), as well as the individuals working within them, with legal or customary authority or recognized responsibilities for managing the heritage in its entirety or parts of it.

3 International Co-Sponsored Meeting on Culture, Heritage and Climate Change (ICSM CHC) held online on December 6–10, 2021.

15.2. Shifting approaches to World Heritage Capacity Building

Considering that capacity building provides important benefits and returns on investment for those involved in the management of World Heritage, in particular for State Parties to the World Heritage Convention, it has been a priority in strengthening the implementation of the Convention for more than two decades. In particular, capacity building was recognized as one of the Strategic Objectives of the World Heritage Convention or 5 C's in 2002 (World Heritage Committee 2002). In replacing the *Global Training Strategy*, the WHCBS proposed some paradigm shifts in the approaches to learning. It highlighted the need to adapt to changes in learning environments and to consider the multiplication of institutions providing capacity building, global access to the world wide web, online access to information and resources and the idea of knowledge acquisition in contrast to the more traditional knowledge transfer. The WHL is building on these shifts and taking them forward as follows.

15.2.1. *From training to capacity building*

The *Global Training Strategy* had already included the idea of looking comprehensively at both natural and cultural heritage, as well as the need to involve two levels of implementation, the international and the regional, following the UNESCO regions (Europe and North America, Asia and the Pacific, Latin America and the Caribbean, Africa and the Arab States). With the WHCBS, the more limited understanding of training, which was focused on fostering knowledge and skills in individual practitioners, is expanded to that of a broader audience of practitioners, institutions, networks and communities. This approach also highlights that building capacities involves a long-term and continuous process which relies on peer-learning and exchange. In particular, it was recognized that the multiplication of training institutions including UNESCO Chairs and Category 2 Centres increased the capacity for delivery. This capacity can now be scaled up with the use of new technologies and in particular the use of online platforms. WHL is consolidating the development of analytical, interpersonal and leadership skills, including communication and management and an interdisciplinary knowledge framework for management planning providing online activities and tools, such as the *Enhancing Our Heritage Toolkit 2.0* (UNESCO et al. forthcoming) and the *Guidance and Toolkit for Impact Assessments in a World Heritage Context* (UNESCO et al. 2022), while cooperating with regional and international research and training institutions, using the PANORAMA Solutions for a Healthy Planet[4] platform as a virtual

4 PANORAMA Solutions for a Healthy Planet is a partnership between 12 organizations, led by IUCN and GIZ (Deutsche Gesellschaft für Internationale Zusammenarbeit) that showcases

repository of case studies and good practices and launching a Managing World Heritage capacity building platform to showcase and continuously update learning resources. The potential of wider outreach beyond World Heritage provides a wider access which supports the development of a community of practice that can be better equipped to collaborate in confronting climate change impacts and in sharing solutions for mitigation and adaptation.

15.2.2. *The heritage place approach*

The WHCBS notes that many management issues are arising from factors occurring outside of the properties and even outside of their buffer zones. This is particularly true for climate change. WHL proposes using a place-based and people-centered approach to management, which promotes the relational concept of heritage place (Franceschini et al. 2022). Such an approach requires the understanding and analysis of the social, environmental, economic and political contexts, namely the property's wider setting, that provides a more in-depth understanding of the origins of the factors impacting the property and buffer zone, enabling not only a deeper understanding of World Heritage properties and their multiple heritage values beyond the outstanding universal value for which they have been inscribed on the list, but also enabling the identification of actors involved, impacting and impacted by the management and conservation of properties, including rightsholders and stakeholders. An emphasis is put on governance, as the heritage place approach helps identify the relevant partnerships in need of being established for fostering sustainable development and the long-term conservation of World Heritage. With this approach, the factors affecting World Heritage properties that are related to climate change can be disentangled, as well as the attributes that are being impacted by these factors, determining the causes and responsibilities to address these holistically and comprehensively, which should also have an influence on governance (re-)arrangements.

15.2.3. *Nature and culture*

The Convention is the only international legal instrument that brings together cultural and natural heritage concerns. Even though it provides a platform of exchange between both fields of practice, separate realms of implementation have been consolidated through time, while different World Heritage actors have continuously addressed the need to find common ground (Von Droste et al. 1998; Larsen and Wijesuriya 2015). In the first decade of implementation of the WHCBS,

---

case studies that address both conservation and sustainable development via an online platform: https://panorama.solutions/en.

initiatives have been undertaken by the Advisory Bodies to increase the exchange and devise potential working methods to break the divide between natural and cultural heritage practices (Leitão and Badman 2014; Leitão et al. 2018; McIntyre-Tamwoy and Badman 2020). One of the results is the partnership of ICCROM and IUCN in delivering the WHL. One of its pillars is to work with natural and cultural heritage practitioners and World Heritage natural and cultural properties together. This approach provides an exchange between sectors and disciplines necessary for addressing climate change concerns, particularly because there are tested methods and tools already in use by the natural heritage sector from which the cultural heritage sector can benefit. One of these relates to management effectiveness assessment with the adaptation of the *Enhancing Our Heritage Toolkit 2.0* (UNESCO et al. forthcoming), which was originally developed for assessing the management effectiveness of natural properties, and now has been developed to also be used for cultural properties, and the *Guidance and Toolkit for Impact Assessments in a World Heritage Context* (UNESCO et al. 2022) that has been produced integrating the methods for Environmental Impact Assessments and Heritage Impact Assessments into one methodology applicable for all World Heritage properties, also in coordination with the wider field of Impact Assessments. In their approach to Indigenous and traditional knowledge as well as local communities, the natural heritage sector can expand its strategies by using perspectives developed by the cultural heritage sector advancing transdisciplinary approaches, which can already be seen in the assessments developed by the Intergovernmental Science-Policy Panel Biodiversity and Ecosystem Services (IPBES) where Indigenous and Local Knowledge is being taken into account in relation to biodiversity conservation and diverse approaches to natural systems.

To bridge the gap and facilitate exchange, a common conceptual framework for management was necessary. The *World Heritage Resource Manual for Managing World Heritage* and the *People Nature Culture* (PNC) course – which has become the flagship management course delivered by the WHL in partnership with different institutions in different regions of the world – looks to establish a common ground with a *Knowledge Framework for Managing World Heritage*, which also includes a *Glossary* that can allow practitioners to speak the same "language". Targeting managers, but also practitioners and members of communities involved in the management of World Heritage properties and other heritage places, the PNC provides participants with a framework that assists them in management planning processes. In particular, with the PNC, managers become equipped with concepts and tools that support them in deepening their knowledge on the heritage values, management system and governance arrangements of their sites, which will allow them to create better management responses. In using this framework, climate change concerns can be integrated into management planning and management effectiveness assessments.

Moreover, the PANORAMA Nature-Culture Community launched in 2020 is a space where natural and cultural heritage practitioners can exchange their work and strategies while sharing with a wider inter-sectoral audience. The Nature-Culture Community has become a repository for the case studies of participants attending the diverse WHL activities enabling managers to reflect on their good practices. In addition, the Site Managers Forum, that is held in parallel to the World Heritage Committee Sessions since 2017, has become a space for World Heritage managers of both sectors to share challenges and management practices and to be updated with the latest resources on the field.

15.2.4. *Practice–Research–Policy Nexus*

Similar to the links between natural and cultural heritage, the links between practice, research and policy have yet to be sufficiently exploited in the World Heritage system. To better address the needs of practice in terms of research and the needs for evidence by policy-makers, the WHL has created an experimental platform, the Heritage Place Lab (World Heritage Leadership 2022; Ishizawa and Jo forthcoming). The aim is to facilitate linkages between research and practice to support World Heritage site management and policy-making by applying a heritage place approach to a research–practice laboratory. The Heritage Place Lab functions as an incubator for practice-led research agendas for World Heritage properties and in its pilot phase has tested the model of the research–practice team, which can be scaled up and used by other World Heritage properties. This model consists of a partnership between research and management institutions working at specific World Heritage sites, which can be foreseen as a long-term collaboration. The premise of the Heritage Place Lab is that working collaboratively between managers and researchers will support a better-targeted research in which results can easily be applied and tested on the ground. Furthermore, the Heritage Place Lab sees research as the nexus between practice and policy, and by reinforcing practice-driven research, the practical needs can be translated into policy more efficiently with the ultimate goal of conserving World Heritage. The 2007 *Policy* already stated the following:

> Research will also be used as a means for capacity building among site managers and to raise awareness among the public which, in turn, may help build public and political support. Involving properties in research relating to climate change will also enable them to identify specific research needs and better apply research results to assist the management of the properties. (UNESCO 2007b, p. 5)

It furthermore mentioned that

> problems being experienced by managers will be clearly translated into research questions to ensure that gaps in knowledge are identified and are used to inform the development of relevant research programmes and translation of such research into useful guidelines and protocols for best practice. (UNESCO 2007b, p. 6)

The Heritage Place Lab, in its pilot phase, proposed climate change as a priority theme and explored with eight World Heritage sites in four world regions, existing climate change-related factors affecting the properties, as well as existing and potential risks, and gaps in knowledge for both researchers and managers. As a result, a number of research–practice teams concluded that climate change is an essential topic to be included in their sites' research agendas and agreed that it is a knowledge area where both researchers and managers need further capacity development to understand impacts on properties and develop adequate management responses (Ishizawa and Jo 2023). In its consolidating phase, the Heritage Place Lab will focus on assessing monitoring systems for the state of conservation of properties and identify climate change research priorities to propose guidance for developing collaborative research agendas for World Heritage sites. In particular, the potential of citizen science and the use of traditional and indigenous knowledges for monitoring the state of conservation of properties, in particular, in relation to factors related to climate change, will be explored. The knowledge has been developed collectively by communities through series of trials and errors based on direct observation of the local environmental context.

15.2.5. *Resilience and disaster risk management*

Based on the yearly review of properties' state of conservation reports and the periodic reporting exercises done at regional levels, the WHCBS identified "Disaster risk reduction" as one of the key themes in need of new tools and methods for improving the work of heritage practitioners. In order to cover this need, several initiatives for education have been promoted in the last decade to develop the capacities of managers and heritage professionals to work on DRM plans for World Heritage properties. Notably, how cultural heritage has been integrated in the disaster risk policy in Japan has been an example promoted by the UNESCO Chair on Cultural Heritage and Risk Management (Jigyasu et al. 2021), the UNESCO Chair on Nature-Culture Linkages in Heritage Conservation (Ishizawa et al. 2019), programs undertaken in cooperation with ICCROM, IUCN, ICOMOS and UNESCO, and the World Bank (Newman et al. 2020).

From the outset, resilience and DRM have been key components of the WHL program. Thus, the WHL is currently working on a new manual on *Managing Disaster Risks and Building Resilience* that will update the previous *Managing Disaster Risks for World Heritage* (UNESCO et al. 2010) integrating climate concerns in a more focused manner. In the updated version of the manual, the impact of climate change on disaster risks and nature-based solutions for adaptation and risk reduction have been explicitly addressed.

DRM helps by suggesting general principles that managers can apply to a specific heritage place and a methodology that emphasizes reducing risk at every step. It aims to prevent new risks, reduce existing risks and manage any residual risks, while strengthening a heritage place's resilience through the continuous improvement in disaster preparedness, response and recovery practices.

While it is recognized that management cannot always eliminate a hazard, there is a lot that can be done proactively to reduce vulnerabilities and enhance resilience so that the disaster risk is lower. There are also approaches that can be taken in the event of a disaster so that the best possible solutions can be found for people and places. Heritage can be part of those solutions, for example, traditional knowledge systems and nature-based solutions can reduce vulnerability and enhance resilience in a way that is place specific.

15.3. Conclusion: a way forward

A complex subject that needs to be holistically tackled, climate change is controversial in the context of an international treaty such as the 1972 Convention which in its Article 4 gives the responsibility for maintaining World Heritage sites for future generations to State Parties. Taking presence at a very slow pace, in 2011, climate change still needed to be highlighted as an emerging theme for capacity building in the WHCBS. It has taken around two decades for climate change to become a forefront issue that World Heritage managers and all actors linked to the implementation of the Convention urgently need to address. To confront this, resources and activities are being designed and implemented by the WHL. However, there is a great limitation to these efforts, as all capacity building activities within the World Heritage Convention including that of WHL are being implemented with the voluntary contributions of select financial partners and institutions. For more concentrated efforts on climate action, more resources and efforts need to be allocated at a general level.

Building on the premises of the WHCBS, the WHL is taking advantage of the potential of wider outreach provided by new technologies to develop a community of practice that can collaborate in confronting climate change issues at World Heritage properties. Using the heritage place approach to management, climate change impacts can be better understood, and management responses can be holistically designed to provide innovative solutions. The *Knowledge Framework for Managing World Heritage* provides a common interdisciplinary framework for natural and cultural heritage practitioners, allowing climate change concerns to be integrated into management planning and management effectiveness assessments.

Activities such as the Heritage Place Lab and the Disaster Risk Management and Resilience courses and workshops put climate change as a central topic, equipping practitioners and institutions with analytical skills and interdisciplinary knowledge that gives managers a deeper understanding of their properties and their multiple values and enables them to connect with disciplines of work adopting an intersectoral approach. From these experiences, the WHL notes that for heritage to address climate challenges, managers and practitioners need to work with other sectors and heritage management planning needs to be embedded in larger frameworks of action. As the 2007 *Policy* pointed out,

> World Heritage properties can serve as laboratories where monitoring, mitigation and adaptation processes can be applied, tested and improved. (UNESCO 2007b, p. 4)

In that sense, WHL is encouraging networking and peer learning by creating joint opportunities for practitioners at global, regional and national levels and platforms for World Heritage properties and the people working on them to connect and explore together. Nevertheless, it is urgent to develop common tools and methodologies that can help site managers to take climate action as part of their management system. The WHL is now embarking with ICOMOS and the UNESCO World Heritage Centre in developing guidance and a toolkit for climate action that responds to the updated Policy to be adopted in 2024. Activities already being disseminated by the WHL will serve as testing grounds to support the integration of climate change concerns in capacity building and taking climate action for World Heritage.

15.4. References

Franceschini, N., Court, S., Ishizawa, M., Jo, E. (2022). Heritage place approach: Linking people, nature and culture in the management of world heritage. In *1972/2022 World Heritage in Transition: About Management, Protection and Sustainability*, De Luca, G., Caccia Gherardini, S., De Vita, M., Francini, C. (eds). Firenze University Press, Florence.

Guzman, P., Fatorić, S., Ishizawa, M. (2020). Monitoring climate change in World Heritage properties: Evaluating landscape-based approach in the state of conservation system. *Climate*, 8(3), 39.

Heron, S.F., Eakin, C.M., Douvere, F. (2017). *Impacts of Climate Change on World Heritage Coral Reefs: A First Global Scientific Assessment*. UNESCO World Heritage Centre, Paris.

ICOMOS (2019). *Future of Our Pasts: Engaging Cultural Heritage in Climate Action.* ICOMOS, Paris.

Ishizawa, M. and Jo, E. (eds) (2023). Towards practice-led research agendas for World Heritage properties. *Journal of Cultural Heritage Management and Sustainable Development*, 13(3), Special issue.

Ishizawa, M. and Jo, E. (eds) (Forthcoming). *Heritage Place Lab, a Model for Research-Practice Collaboration in the Context of World Heritage*. ICCROM, Rome.

Ishizawa, M., Inaba, N., Yoshida, M. (eds) (2019). *Proceedings of the Third Capacity Building Workshop on Nature-Culture Linkages in Heritage Conservation in Asia and the Pacific 2018: Disasters and Resilience*. September 21–October 1, 2018, Tsukuba.

Jigyasu, R., Kim, D., Shakya, L., Cheek, W. (2021). *Proceedings of UNESCO Chair Programme on Cultural Heritage and Risk Management*, Ritsumeikan University, Kyoto.

Larsen, P.B. and Wijesuriya, G. (2015). Nature-culture interlinkages in World Heritage. Bridging the gap. *World Heritage*, 75, 4–15.

Leissner, J. and Kilian, R. (eds) (2014). *Climate for Culture. Built Cultural Heritage in Times of Climate Change*. Fraunhofer Moez, Leipzig.

Leitão, L. and Badman, T. (2014). Opportunities for integration of cultural and natural heritage perspectives under the World Heritage convention. Towards connected practice. In *Conserving Cultural Landscapes. Challenges and New Directions*, Taylor, K., St. Clair, A., Mitchell, N.J. (eds). Routledge, New York.

McIntyre-Tamwoy, S. and Badman, T. (2020). The culture nature journey from Honolulu, Hawaii, to Sydney, Australia: Reaching agreement on what's next? *Historic Environment*, 32(1), 70–86.

Morel, H., Megarry, W., Potts, A., Hosagrahar, J., Roberts, D.C., Arikan, Y., Brondizio, E., Cassar, M., Flato, G., Forgesson, S. et al. (2022). *Global Research and Action Agenda on Culture, Heritage and Climate Change.* ICOMOS & ICSM CHC, Charenton-le-Pont and Paris.

Newman, J.P., Minguez Garcia, B., Kawakami, K., Akieda, Y., Naito, I. (2020). *Resilient Cultural Heritage: Learning from the Japanese Experience*. World Bank Group, Washington.

Orlove, B., Dawson, N., Sherpa, P., Adelekan, I., Alangui, W., Carmona, R., Coen, D., Nelson, M., Reyes-Garcia, V., Rubis, J. et al. (2022). *ICSM CHC White Paper I: Intangible Cultural Heritage, Diverse Knowledge Systems and Climate Change: Contribution of Knowledge Systems Group I to the International Co-Sponsored Meeting on Culture, Heritage and Climate Change*. ICOMOS & ICSM CHC, Charenton-le-Pont and Paris.

Seekamp, E. and Jo, E. (2020). Resilience and transformation of heritage sites to accommodate for loss and learning in a changing climate. *Climate Change*, 162(1), 41–55.

Shepherd, N., Cohen, J.B., Carmen, W., Chundu, M., Ernsten, C., Guevara, O., Haas, F., Hussain, S.T., Riede, F., Siders, A.R. et al. (2022). *ICSM CHC White Paper III: The Role of Cultural and Natural Heritage for Climate Action: Contribution of Solutions Group III to the International Co-Sponsored Meeting on Culture, Heritage and Climate Change*. ICOMOS & ICSM CHC, Charenton-le-Pont and Paris.

Simpson, N.P., Orr, S.A., Sabour, S., Clarke, J., Ishizawa, M., Feener, M., Ballard, C., Mascarenhas, P.V., Pinho, P., Bosson, J.B. et al. (2022). *ICSM CHC White Paper II: Impacts, Vulnerability, and Understanding Risks of Climate Change for Culture and Heritage: Contribution of Impacts Group II to the International Co-Sponsored Meeting on Culture, Heritage and Climate Change*. ICOMOS & ICSM CHC, Charenton-le-Pont and Paris.

UNESCO (2007a). Climate change and World Heritage: Report on predicting and managing the impacts of climate change on World Heritage and strategy to assist State parties to implement appropriate management responses. *World Heritage Reports Series 22*, UNESCO, Paris.

UNESCO (2007b). Policy on the impacts of climate change on World Heritage. Policy document, UNESCO, Paris.

UNESCO (2007c). *Case Studies on Climate Change and World Heritage*. UNESCO, Paris.

UNESCO (2011). World Heritage capacity building strategy. Document, UNESCO, Paris.

UNESCO (2014). Climate change adaptation for natural World Heritage sites: A practical guide. *World Heritage Papers Series 37*, UNESCO, Paris

UNESCO (2021). Updating of the policy document on the impacts of climate change on World Heritage properties (WHC/21/44.COM/7C). UNESCO, Paris.

UNESCO, ICCROM, ICOMOS, IUCN (2010). Managing disaster risks for World Heritage. UNESCO, Paris.

UNESCO, WRI, IUCN (2021). World Heritage forests: Carbon sinks under pressure. Report, UNESCO, Paris; WRI, Washington; IUCN, Gland.

UNESCO, ICCROM, ICOMOS, IUCN (2022). *Guidance and Toolkit for Impact Assessments in a World Heritage Context*. UNESCO, Paris.

UNESCO, IUCN, ICCROM, ICOMOS (Forthcoming). *Enhancing Our Heritage Toolkit 2.0. Assessing Management Effectiveness of World Heritage properties*. UNESCO, Paris.

Von Droste, B., Rössler, M., Titchen, S. (eds) (1998). Linking nature and culture. Report, Global Strategy Natural and Cultural Heritage Expert Meeting, 25 to 29 March 1998, Amsterdam; UNESCO World Heritage Centre, Paris.

World Heritage Committee (2001). Progress report on Global Training Strategy (WHC-01/CONF.208/14). Report, UNESCO, Paris.

World Heritage Committee (2002). Budapest declaration on World Heritage (WHC-02/CONF.202/5). Report, UNESCO, Paris.

World Heritage Committee (2005). Decisions of the 29th session of the World Heritage Committee (Durban, 2005). Report, UNESCO, Paris.

World Heritage Leadership (2022). Heritage Place Lab, interlinking research and practice for improving World Heritage management. PANORAMA Solutions for a Healthy Planet [Online]. Available at: https://panorama.solutions/en/solution/heritage-place-lab-interlinking-research-and-practice-improving-world-heritage-management.

APPENDICES

Reflections on the Future of Heritage Education

Appendix 1

Heritage as a Discipline in the French Schools of Architecture

In France, the regulatory framework governing the training curricula of architectural studies leading to the award of the bachelor's degree (*licence*) in architectural studies and the state architect's diploma conferring a master's degree is defined by the Decree of July 20, 2005. This decree recognizes the significance of heritage studies in the curricula of schools of architecture, encompassing a broad spectrum of disciplines ranging from the preservation of historic monuments to the refurbishment of existing edifices. The field of architectural and urban heritage intervention holds great potential as a future market, prompting a widespread acknowledgment of the need for enhanced training among architects and other stakeholders involved in the architectural and urban milieu. Accordingly, it is imperative that architects receive comprehensive and rigorous education in order to devise tailored solutions that cater to the diverse demands of their clients.

The multidimensional aspect of architectural studies is acknowledged by its various disciplines and teaching methods, which have evolved to incorporate this perspective. As per the ministerial decree, schools of architecture are mandated to integrate this aspect into their curriculum, and thereby facilitate the acquisition of a comprehensive understanding of existing buildings by imparting lessons in the history of architecture and urban planning, surveying techniques, constructive elements and regulatory tools relevant to this domain. The educational institutions have also prioritized the development of competencies in different modes of intervention pertaining to existing structures, which includes preservation, restoration, refurbishment and adaptive reuse.

Chapter written by Lazaros MAVROMATIDIS.

Furthermore, the significance of this dimension is underscored by the fact that it fosters an appreciation for the cultural and societal values embodied by historic edifices, while simultaneously promoting sustainable practices in contemporary construction. Consequently, the pedagogical approach adopted by architectural schools has shifted to a more holistic and interdisciplinary approach, which enables students to develop a nuanced perspective that considers the broader social, economic and environmental implications of their designs. The goal is to create graduates who possess a deep understanding of the complexities involved in the preservation and enhancement of existing architectural and urban heritage, and who are equipped to apply this knowledge in a practical setting.

A study conducted by the National Council of the Order of Architects in 2014 revealed that intervention in architectural heritage constituted a significant portion of the building market, despite only accounting for a third of the activities carried out by architects. It is worth noting that the scope of this market extends beyond architects and involves the collaboration of a diverse range of professionals such as engineers, technical design offices, construction companies, and public or private project management consultancies. The field of intervention in heritage encompasses a wide range of activities at varying spatial scales, spanning from minor improvements to extensive modifications of the built environment. As a result, it is a complex field that necessitates specialized technical, methodological and regulatory skills to execute such interventions successfully.

It is imperative to recognize that this sector is multi-faceted, involving diverse actors who work at multiple levels of the design and construction process. Consequently, interventions in architectural heritage demand a collaborative approach that incorporates expertise from various domains, such as conservation, restoration and adaptive reuse. To succeed in this field, professionals must possess a comprehensive understanding of the regulatory framework, technical requirements and innovative methods for the preservation and enhancement of architectural heritage. Such expertise is vital to ensure that the interventions are sustainable, compatible with the cultural and social context of the region, and aligned with the evolving needs of the society.

A thorough analysis of the curricula of numerous French schools of architecture revealed a concerning lack of emphasis on teaching programs related to architectural heritage. Despite the fact that this field constitutes a significant portion of the building market, these schools still predominantly focus on new construction projects. Moreover, there seems to be a disconnect between the teaching of cultural heritage and climate action, as the former is not adequately integrated into the pedagogical innovations aimed at climate preservation. The underwhelming interest of architects in this market, coupled with the emergence of specialized technical project management roles, necessitates a re-evaluation of the structure and content of the teaching programs.

This report endeavors to identify actionable strategies that can enhance the resiliency of architectural heritage while creating meaningful connections between this domain and climate preservation initiatives. To this end, we examined the training programs available to key actors involved in rehabilitation, including architects, project managers and other related professions. Our analysis identified various potential avenues for pedagogical interventions that take into account the unique characteristics of the French pedagogical landscape.

In order to arrive at our conclusions, we scrutinized the teachings of diverse institutions such as Schools of Architecture, Schools of Applied Arts, Universities, University Institutes of Technology and continuing education providers. Our goal was to ascertain the strengths and limitations of the existing training programs and to identify opportunities for developing innovative pedagogical approaches that could equip future generations of professionals with the skills and knowledge required to address the complex challenges of preserving architectural heritage in a changing world.

It is important to note that the process of identifying relevant information on educational programs proved to be a challenging task due to their limited availability on the websites of various Schools. Additionally, documenting the pedagogical outcomes of diverse training courses proved to be even more complex, given the scarcity of publications featuring student works, the absence of a reliable archiving policy for the results in media libraries of the Schools and the limited opportunities for exchanging ideas between teachers, which are usually restricted to workshops and Final Diploma Project juries. The outcome of the state of the art study was a comprehensive identification of the main trends and specificities observed in the educational institutions that are involved in heritage interventions, along with an analysis of the various study days that were dedicated to the relationship between architectural projects and heritage. This chapter aims to provide a concise summary of the study's findings, which is not intended to serve as a qualitative evaluation of the curriculum or a quantitative assessment of the dominant teaching narrative pertaining to rehabilitation in the context of heritage. Despite these challenges, the study was able to identify several emerging educational trends and innovations that are gaining traction in schools of architecture. These promising developments could potentially serve as the basis for further exploration and experimentation in the realm of architectural heritage, as educational institutions work toward equipping future professionals with the tools, skills and knowledge necessary to tackle the multifaceted challenges posed by the preservation and restoration of heritage sites.

A1.1. The teaching of "*heritage*" in the form of "*rehabilitation*" in France

In the French educational landscape, the built environment is an interdisciplinary field that is approached through diverse lenses and objectives. In the context of architectural design studios, project teachers perceive existing buildings as a reservoir of architectural potential, the recognition of which quality and transformation must be addressed through the culture of the project. However, despite the potential synergies, heritage and climate change are rarely addressed in a trans-disciplinary manner across the different pedagogical practices. For history teachers, these architectures represent a corpus that necessitates a critical appraisal, focusing on their morphological characteristics, especially in terms of rehabilitation which aims to restore the edifices' original qualities and enable reversibility. For construction teachers, these buildings are a synthesis of *utilitas*, *firmitas* and *de venustas* values, particularly for the architectures of the 20th century, where modern materials like the innovative reinforced concrete were extensively used, thus making these edifices a vehicle of innovation.

The ministerial directives have established three fundamental postulates to encapsulate the disciplinary knowledge pertaining to the rehabilitation of existing built environments. The first postulate posits that the field of action extends beyond the confines of the institutionally protected heritage and encompasses the entire gamut of the *already there*. The second postulate is premised on the creative dimension of rehabilitation, which necessitates a *projectual* approach that goes beyond mere technical or regulatory responses. The existing built environment is not merely a sector of the construction market, but is also a repository of immense cultural potential. Lastly, the third postulate underscores the operational skills that architects must possess, which include the ability to recognise the architectural identity of a building (its unique history, formal and typological values, and inherent qualities) and to comprehend its materiality (its constructive structure, materials and pathologies). These postulates serve as the foundation of an integrative approach to rehabilitation that fosters a deep understanding of the built environment, its cultural significance and its technical exigencies.

For instance, the first postulate implies that the scope of intervention extends beyond buildings and sites that are legally protected to encompass other types of buildings and landscapes that are significant and worth preserving. This recognizes that heritage is not just about the exceptional or monumental, but also about the ordinary and everyday, which constitute a collective memory and cultural identity. The second postulate highlights the need for a creative and project-oriented approach to rehabilitation that goes beyond a purely technical or regulatory response. This means taking into account the broader context, the stakeholders involved, the social and cultural dimensions, and the environmental impacts. The third postulate emphasizes the importance of operational skills that are specific to heritage rehabilitation, which include the ability to identify and assess the architectural identity of a building, as well as its materiality and pathologies. These skills require a deep understanding of architectural history, construction techniques and materials science, as well as the ability to work collaboratively with other professionals and stakeholders.

Overall, these postulates reflect a comprehensive and multidisciplinary approach to heritage rehabilitation that takes into account not only the technical and regulatory aspects, but also the cultural, social and environmental dimensions. They provide a framework for architects and other professionals to engage with heritage rehabilitation in a way that is both respectful of the past and relevant for the present and future.

The aforementioned postulates are established guidelines that serve as a basis for various educational institutions to cultivate a pedagogical environment that imparts an appreciation of the built environment and aims to cultivate a certain degree of awareness of the cultural and historical value that it represents. The level of emphasis placed on each of these postulates varies, depending on the institutional framework of the educational entity in question. Some university faculties that offer master's courses such as policy-making and geography exhibit a clear commitment to these postulates, as they provide their students with a comprehensive understanding of the interrelationship between the built environment, heritage and climate change.

The architecture education system in France is designed in such a way that it grants each school a certain degree of autonomy in terms of the structure and content of their curricula. This autonomy allows the individual schools to develop unique curricula that are aligned with their particular character, and deepen their ties to the local environment while leveraging the expertise of their faculty members. This flexibility also promotes innovation in teaching methodologies, which ultimately enhances the quality of education being offered.

The consensus among architecture educators who are passionate about heritage is that rehabilitation training must not be regarded as a mere specialization, but rather as an integral part of the general education of architects, which should be imparted from the bachelor's degree to the master's level. The urgent need to address the changing urban landscape and the imperative of mitigating the effects of climate change has underscored the importance of reinforcing this type of education in schools of architecture. Furthermore, the concept of heritage has evolved and now extends beyond mere conservation of historic monuments. It encompasses a re-evaluation of the existing built environment through contemporary interventions, programmatic changes and even interventions at the urban level.

Furthermore, there is a growing awareness of the need for architects to be equipped with multidisciplinary skills to address complex issues related to the built environment. The recognition of the interdependence of urban planning, environmental issues, social and cultural aspects and economic factors has led to an increased interest in trans-disciplinary approaches to architectural education. This involves collaborating with other disciplines such as engineering, sociology, anthropology and environmental studies to create a more holistic understanding of the built environment and the challenges it faces. In addition, the pedagogical approaches used in schools of architecture have evolved in response to these changing needs. There is a growing emphasis on project-based learning, which involves students working on real-world projects and engaging with stakeholders, such as community groups and local authorities. This approach allows students to develop practical skills, gain experience in working collaboratively and become familiar with the complexities of the built environment. Overall, the evolution of architectural education in France reflects the changing needs and priorities of society. The integration of heritage and sustainability issues into the curriculum, the emphasis on trans-disciplinary approaches and the use of project-based learning all contribute to the development of a new generation of architects who are equipped to address the complex challenges facing the built environment.

NOTE.– The training of architects in the intervention on existing buildings constitutes a field of education which gradually asserted itself in France in the last years of the 20th century, in a context of growth of the rehabilitation market. This progression is the result of a considerable effort by the public authorities and professional organizations to transform, in the perception of architects, the status of maintenance and improvement operations of the existing building stock into operations of architectural creation to the architectural profession. This objective appears clearly in the general structure of the teaching of architecture developed during the reform of the teaching of the Schools of Architecture in 2005. This reform introduces the BMD university system (bachelor's, master's, doctorate) and organizes studies in architecture over 5 years, articulated in two cycles of 3 and 2 years, bachelor's and master's. These can be followed by a third cycle of doctorate, conditional on the presence in the Schools of accredited research laboratories associated with a university doctoral school. If the first 3-year cycle concerns fundamental lessons, it is the second cycle which institutes the lessons of specialized knowledge.

Among the latter, that of the rehabilitation of the built heritage constitutes a field of study which has gradually acquired a real status. It was during the 2000s that the debate on the diversification of architectural professions prompted the Schools to gradually offer training at master's and post-master's level in professional specialisation in the rehabilitation sector. It was a question of opening up the field of activity of architects more widely to intervention on existing buildings and responding to a desire for heritage recognition of recent buildings and sites, repositories of a value of memory, but not concerned by the protective measures in respect of historic monuments. This objective has been present since the 2005 reform, which indicates in particular that the emphasis must be placed, on the one hand, on sustainable development and, on the other hand, on the rehabilitation and interventions on existing buildings, in other words, heritage in its broadest sense. Appendix 2 of the text of the reform indicates that

> … the various actors intervening in an architectural and urban context, and even more the architects, must be able to be better trained in order to provide responses adapted to the diversity of orders. The confrontation of ancient and contemporary architecture remains one of the most constant situations in architectural design... Work on what already exists – both old and more recent – must make an integrated project approach possible. Within architectural studies, the various disciplines and the teaching of the project will take this dimension into account. The Schools of Architecture will contribute, throughout their teaching, to a better understanding of existing buildings, by developing lessons in the history of architecture and town planning, surveying techniques, construction methods and regulatory tools concerning it. They will focus on learning the different modes of intervention on what already exists (author's translation).

These objectives have been validated by the various ministerial reports, in particular those of the Ministry of Culture and Communication, which prepared the reform of the teaching of the Schools of Architecture published in February 2018.

A1.2. Perspectives of evolution focusing on the French academic environment

Upon conducting extensive research on the curricula of various schools of architecture in France, I have uncovered noteworthy pedagogical innovations and untapped teaching potentials. One of the key areas that require attention pertains to the need for the systematic expansion of heritage education with respect to climate change in the undergraduate cycle. To accomplish this, it is necessary for French schools of architecture to establish a shared frame of reference that places emphasis on the intersection of heritage and climate change in the master's cycle through the implementation of specific fields of study. Furthermore, it is essential to develop a progressive and complementary approach to teaching by leveraging innovative pedagogical strategies.

It is worth noting that the integration of climate change concerns in the context of heritage education represents a pressing matter that demands swift and decisive action. By broadening the scope of heritage education to include climate change-related considerations at the undergraduate level, students will be better equipped to navigate the complexities of the rapidly evolving socio-environmental landscape. Moreover, the establishment of a shared frame of reference focusing on heritage and climate change in the master's cycle would provide students with the necessary tools and knowledge to tackle these issues from a multi-disciplinary perspective. This is particularly important given the urgent need for effective and sustainable solutions to mitigate the adverse effects of climate change on the built environment.

In order to achieve the aforementioned objectives, it is imperative that schools of architecture in France prioritize the development and implementation of innovative pedagogical actions that promote progressiveness and complementarity in teaching. This involves leveraging emerging technologies, creative teaching methodologies and interdisciplinary collaborations to foster a more dynamic and engaging learning experience. Ultimately, this approach will enable students to develop a more nuanced and sophisticated understanding of heritage and climate change, and empower them to become agents of positive change in the built environment.

To optimize the management of curricula, it is essential to introduce a dynamic and flexible approach that entails a partial pooling of general and specialized education, and even inter-school education, as well as establishing a balance between theoretical and practical knowledge. Moreover, to ensure a comprehensive approach, it is crucial to extend the concept of heritage beyond traditional notions, such as monumental heritage, to include undistinguished architectures, rural landscapes and industrial heritage. Tutors should also consider the project approach in conjunction with a theoretical posture and a scientific methodology, with the integration of engineering science to anticipate climate change effects. This interdisciplinary approach will aid in the introduction of research methodologies and the integration of technical contingencies that are subjected to perpetual evolution into teaching. Another pressing issue is the inclusion of interdisciplinary research on climate change within the architectural design studios that focus on heritage, as this will enable students to gain a better understanding of the implications of climate change on existing built environments.

Additionally, the introduction of new technologies into the heritage education process could enhance teaching effectiveness and reinforce the theoretical and scientific dimensions of the courses. Using virtual and augmented reality, building information modeling (BIM), geographic information systems (GIS) and other innovative tools could improve the ability of students to analyze, model and simulate different scenarios related to the rehabilitation of heritage buildings in the context of climate change. This would enable students to explore and experiment with various design options and their environmental impact, promoting a more sustainable and resilient approach to architectural heritage.

In conclusion, the development of a comprehensive and innovative pedagogy in heritage education within schools of architecture in France is crucial for addressing the complex challenges posed by climate change and for promoting a more sustainable and resilient approach to architectural heritage. The incorporation of interdisciplinary and technical elements, as well as the use of innovative technologies, could enhance the effectiveness of the teaching process and enable students to develop the skills and knowledge needed to address the challenges of the future.

The importance of openness in pedagogical praxis cannot be overstated. Specifically, the incorporation of real-world professional experiences and engagement with the regulatory framework of the European Union can greatly enrich the education of aspiring architects. This approach not only serves to sensitize students to contemporary societal issues and EU initiatives, but also cultivates a more global perspective through fostering international partnerships with foreign institutions. In particular, inter-school thematic collaborations offer a valuable opportunity for schools of architecture to pool resources and expertise in the design and implementation of complex projects. Ultimately, such endeavors can generate a fertile ground for the cultivation of innovative pedagogical approaches and the promotion of sustainable architectural practices.

In architectural teaching programs, it is imperative that the rehabilitation of existing buildings be approached as a highly complex endeavor, far surpassing the realm of mere improvement or technical updating. Instead, tutors must engage with the architectural project in a deeply informed manner, one which takes into account the unique history of the existing building and its various spatial, structural and mnemonic qualities. In order to cultivate a truly cultural approach and fully realise the potential of the existing structure as a foundational element of the project, relevant courses should be incorporated into the curriculum. Such courses ought to place emphasis on the construction of meaning and the exploration of the material potentialities of the existing structure, whether the project calls for restoration, restitution, recovery of materials or restructuring. To this end, the centrality of the project workshop should be affirmed, with particular attention paid to the exploration of diverse pedagogical postures.

It is essential to abandon the dichotomy between the sporadic introduction of isolated teachings, motivated by the interests of individual teachers, and the provision of highly competitive, standardized teachings with a perennial structure, aligned with the goal of specialization but lacking the objective of genuine cultural enrichment. To achieve this, it is crucial to integrate heritage courses as the centerpiece of the curriculum, aimed at equipping future architects with a comprehensive set of methodological, technical and theoretical competencies. This would encompass the acquisition of critical skills in the areas of surveying, diagnosis and a deep understanding of the historical, material and immaterial dimensions of heritage.

To further foster the effectiveness of such an approach, the curriculum should also prioritize the development of transversal competencies and encourage cross-disciplinary collaborations. By establishing connections between heritage courses and other disciplines such as geography, sociology, environmental science or economics, students would be able to gain a more comprehensive understanding of the complex issues that surround heritage and its relation to contemporary society. Additionally, such an approach would foster innovation in teaching practices by enabling the development of joint research programs, collaborative workshops or international exchanges with partner institutions. Ultimately, this would lead to a more comprehensive, integrated and effective educational approach that empowers future architects to confront the challenges of heritage management with confidence, creativity and expertise.

As we contemplate the future of architectural education, let us not forget the richness and depth of our past. Let us embrace the notion of heritage as a living, breathing entity that connects us to our roots and informs our vision for the future. Let us see in every building, every structure and every landscape a story waiting to be told and a legacy waiting to be preserved. Let us approach each project with humility, curiosity and an unwavering commitment to excellence. And let us, in the spirit of true innovation, weave together the threads of tradition and modernity to create a tapestry of architectural brilliance that will inspire and delight generations to come.

In the end, the teaching of heritage in architecture is not only about preserving the past, but also about shaping the future. It is about understanding the history and memory of a place and using that knowledge to create something new, something that both reflects the past and looks toward the future. As the great poet T.S. Eliot once said, *The historical sense involves a perception, not only of the pastness of the past, but of its presence.* This idea of the past being ever-present is at the heart of the teaching of heritage in architecture, and it is through this understanding that we can create buildings and spaces that are not just functional, but also meaningful and enduring.

Appendix 2

Climate, Heritage, Resilience

Climate resilient heritage is a new topic in the Netherlands. Although the current number of courses is small, it can be noticed that there is a slowly increasing interest toward this subject in the education of future architects, planners and designers. The analysis of the various curricula in the Dutch Higher Education Institutions (HEIs) shows seven different approaches to the topic:

– Technology-science schools focus mainly on heritage protection/conservation techniques, heritage history and assessment of heritage values. Adaptation methods/strategies are key subjects in both optional and mandatory courses.

– Design studio-based courses approach "Climate resilience heritage" on an irregular and sporadic basis.

– Nature-science schools have a clear focus on climate change, especially mitigation and adaptation issues (nature-based solutions), but without specific relation to natural/built heritage.

– Practice-based schools (academies) do not have mandatory or optional courses related to heritage and climate change. However, design studio-based courses are widely increasing on this topic.

– Humanities-related schools focus on climate adaptation governance, with no specific attention to heritage.

– The design-based studio courses usually work on a case area and conduct fieldwork/workshop in the dedicated area to develop and test adaptation and mitigation strategies/design solutions. The courses that focus more on heritage conservation and value assessment conduct lectures and seminars.

Chapter written by Irene G. CURULLI and Deniz IKIZ KAYA.

– Almost all of the related courses in the Netherlands are offered in master's programs addressing master's students specifically. This shows that such specialization is not provided at the bachelor's level but only at the master's level.

Currently, there is only one Massive Open Online Courses (MOOC) course (providing a certificate) related to industrial heritage transformation, port cities and climate change. There are more climate-resilient heritage-focused MOOC courses coming up in the near future.

These different approaches and limited courses on heritage and climate change in the Dutch HEIs reveal (1) the lack of a multidisciplinary approach to embrace the complexity of this subject; (2) heritage and climate change is not yet considered as specialist education, to also be trained vocationally, but as a component of the general knowledge/profile of an architecture graduate; (3) this topic is addressed at the master's level, targeting students with specialized skills and interests.

It is important to highlight that the climate change adaptation policy of the Dutch government does not explicitly address climate change impacts on cultural heritage (Fatorić and Biesbroek 2020); thus, it does not give any emphasis on special qualifications/expertise required in the architectural profession. We are left to wonder how or if changes in this policy might inform the curriculum in architectural education. Yet, experts often recognize the synergistic relationship between heritage conservation, environmental sustainability and climate change actions. Hence, the rising interest toward climate resilient heritage can be beneficial for both the heritage and climate change experts.

The existing courses/syllabi show four clear points of reflection when planning future academic courses:

– incorporate in courses the correlation between heritage and climate change;

– raise awareness on the potential of cultural heritage to achieve climate change;

– inform students on assessment methods of vulnerability and the impact of climate change on cultural heritage (for prevention);

– familiarize with resilience concepts for mitigation and adaptation strategies (design).

Reference

Fatorić, S. and Biesbroek, R. (2020). Adapting cultural heritage to climate change impacts in the Netherlands: Barriers, interdependences, and strategies for overcoming them. *Climate Change*, 162, 301–320.

Appendix 3

Reflection Paper on Degree and Master's Programs on Heritage and Resilience

Among the many presence-taught Higher Education Study Programmes that were taken into consideration in the scope of the e-Creha project, the team of the UNIMOL selected four and described them according to the provided template. Classifications are taken from the Bologna Process. Two of the shortlisted are regular "Master's Degrees" (MD): a 2-year study program after the bachelor's degree. The other is a "Professional Master's" (Master), which means a 1-year study program that gives specific and professional tools for the enrolled and is usually placed after the master's degree or bachelor's degree; this degree level is particularly common in Italy.

Furthermore, there is a cultural institution that occasionally works as a body of education providing lectures, events and classes. It was included for its commitment to Heritage Conservation and Risk Reduction.

A3.1. University of Bologna – science for the conservation-restoration of cultural heritage (MD) (LM-11)

Here is an International Degree Programme aimed at education and training in chemistry, physics, geology and informatics applied to cultural heritage and conservation taking place in Ravenna. Among the classes, it is also possible to find the

---

Chapter written by Rossella NOCERA, Luciano DE BONIS and Giovanni PARISANI.

following subjects: "air pollution chemistry"; "climate change and climate actions"; "environmental impact on materials, deterioration and aging".

A3.2. University of Camerino – Corso di Laurea Magistrale in Architettura (MD) (LM-04)

This is a 2-year program of 120 credits and structured in single-subject teaching, theme studios, workshops and seminars. It is divided into three topic clusters: CLIMATE, HERITAGE and METROPOLIS, which are described as the most important issues the culture of design is called to face in our time.

A3.3. Master's in Gestione e Conservazione del Patrimonio Mondiale (Professional Master's)

This Professional Master's is aimed at those who wish to acquire the tools and knowledge necessary for the identification, protection, enhancement and communication of the cultural and natural World Heritage, with a particular focus on management. The study plan contains references to Heritage Studies (laws, restoration techniques, archaeology) and references to Resilience ("Risk Assessment on World Heritage" and "Climate Change and its impact on cultural and human landscape"). It is given by a private training body: "Istituto per l'Arte e il Restauro – Palazzo Spinelli" that provides 1500 hours partly online and partly in person in Florence.

A3.4. Polytechnic of Turin – master's in "methods and techniques for governing resilient cities: towards integrated risk management" (Professional Master's)

This 1-year master's program is explicitly oriented toward Resilient Planning and Design and all related issues. It is developed by the Interdepartmental Responsible Risk Resilience Centre (R3C) of the Politecnico di Torino for the analysis of the vulnerability of territories and the evaluation of the urban system's capacity to respond to natural and human risks in a resilience and sustainability perspective. "In a world increasingly affected by significant challenges, it is becoming more important to provide effective solutions to face the effects of environmental, climatic and socio-economic changes. This situation needs to get the advantage of all new capacities in dealing with such profound changes". One of the classes has the name of "Heritage, Risk and Resilience".

A3.5. International Centre for the Study and the Preservation and Restoration of Cultural Property (ICCROM)

Aiming at building capacities for effective risk management of cultural heritage. The aim is to reinforce policies, institutions and local communities in risk-prone regions. Activities of the program include workshops, training materials and tools that promote better integration of cultural heritage into wider disaster risk policies at regional, national and local levels. Moreover, given ICCROM's strong focus on disaster preparedness, it has leveraged its long-standing cooperation with Japan to bring visibility to Italy's rescue of damaged heritage following the earthquakes in Central Italy.

Appendix 4

Climate Change and Heritage Issues in Architecture Curricula: The Cases of Turkey, Finland, Denmark and MOOCs

This text is a reflection paper on the research conducted by the TOBB University of Economics and Technology (TOBB ETU) as part of the O1 Report of the e-CREHA project entitled "Cross-disciplinary pedagogies".

The impact of climate change on heritage needs to be an integral part of architectural education. Knowledge, expertise and experience from different fields and disciplines must be integrated to address climate change as a threat and to embrace heritage as a resource and driver for sustainable development. This situation requires a focus on climate change and heritage, a review of architectural curricula to assess deficiencies and the implementation of a new course in architectural education. This study explores how existing architectural curricula respond to climate change and heritage issues and what pedagogies are used in educational programs. TOBB ETU initially focused on architectural schools in Turkey and conducted a survey to obtain data for courses in Finland, Denmark and Massive Open Online Courses (MOOCs). The survey was conducted between February and November 2021. We collected data that examined the scope, method and content of the courses in the educational programs.

The research had several limitations. An important issue in architecture curricula is the lack of courses at the intersection of heritage and climate change issues. In Turkey, it was difficult to obtain adequate data for quantitative and qualitative analysis due to the insufficient number of courses reached during the survey. There

---

Chapter written by Pelin GÜROL-ÖNGÖREN, Sibel ACAR, Aslı ÖZBEK, Burçin YILMAZ and Irem KÜÇÜK.

was also a language barrier that made it difficult to obtain data. Internet research was conducted in Turkish and English. That is why the content written in local languages could not be included in this survey.

A4.1. The courses in Turkey

There are only two specialized courses on climate and heritage issues in architecture schools in Turkey. One of these courses is "World Heritage and Site Management", which has been offered in Turkish as an elective course at the graduate level in the Department of Architecture at Kadir Has University since 2020. It deals with the impact of climate change on world heritage sites and ways to deal with this. It is noteworthy that this course includes both climate change and heritage. During the 14-week course, this topic is discussed for 1 week. A second course, "Climate Change and Design", has been offered in English by the Faculty of Architecture at Eskişehir Osmangazi University at the graduate level since 2018. It deals with the issue of design in the urban environment in terms of climate resilience. However, there is no definition in the content that deals with the relationship between climate and heritage issues.

A4.2. The courses in Finland

We could not find a course that specifically addressed this issue. We received three different types of data. The first is the summer school. "Climate Change, Health and Architecture", which has been offered online since 2020 as a 2-week intensive program at the graduate level at Aalto University, deals with the impact of climate change on the built environment through future scenarios. It focuses on design and construction by considering mitigation and adaptation strategies.

The second one, created within the framework of the Erasmus+ Project "Digital Climate Change Curriculum for Architectural Education", is carried out with five partners, coordinated by Tampere University of Finland and partners from TU Dublin-Ireland, Aarhus School of Architecture-Denmark, Bologna University-Italy and Taltech University-Estonia. According to the project report published in 2020, a broad framework around the relationship between climate and architecture is drawn and innovative digital pedagogies are also taken as a subject of research. The project is the development of a training module for the evaluation of the relationship between climate and cultural heritage issues in education. The third is a "Doctoral Program for Architecture 2014–2017 Sustainable Built Environment" offered at the Oulu School of Architecture. This program focuses on the relationship with climate change in the Arctic and Northern built environment.

A4.3. The courses in Denmark

There are a few examples in Denmark. The Architecture and Design Institute's master's program "Spatial Design" at the Royal Danish Academy focuses on the design of the built environment. Although the program does not offer a course dealing specifically with heritage and climate issues, the thesis "Education Center for Climate Change" was carried out in this institution in 2020. Another example is the Radical Sustainable Architecture program offered at the Aarhus School of Architecture, which includes two studio courses. Architectural approaches to societal, climatic and environmental challenges are the subject of the program, which explores "transformation", "habitation" and "sustainability". With a focus on creating a resilient and healthy environment, the program also covers history, culture and environmental issues, as well as ecology and climate/climate change issues.

A4.4. Online courses (MOOCs)

Two courses were found on Coursera that address climate and heritage issues in the survey of online education platforms. The course "Planning for Climate Change in African Cities" takes an interdisciplinary approach to the situation of African cities against climate change. During the course, experts from different disciplines discuss the issues of resilience, vulnerability, mitigation and adaptation.

Another course, "Heritage under Threat?" focuses on threatened heritage and engaging global communities. In addition to climate change, the course content includes discussions of other heritage-threatening issues such as war, migration and financial crises. Although the focus of the first course is on climate change and the focus of the second course is on heritage, the relationship between the two is also discussed. As with university programs and courses, the number of courses in the digital environment is quite limited.

A4.5. Pros and cons of addressing climate change and heritage issues together

Architectural education should put climate change and heritage issues on the agenda and find ways to address them in their educational programs. The study found that these issues are mostly discussed separately, and the number of courses referring to them is quite low. A survey of the courses accessed was searched and evaluated according to their content, level of training, years taught, method and training medium. In light of the data obtained, the pros and cons were identified by examining courses in Turkey, Finland, Denmark and MOOCs.

The advantages can be summarized as follows:

– Heritage and climate change are already addressed to some extent in undergraduate and graduate courses. The scope and methods of these courses can be developed.

– Higher Education Institutions (HEIs) can develop online courses to reach a wider audience for dissemination.

– Courses taught in English provide an opportunity for further dissemination of course content and outputs.

– Online courses for lifelong learning are suitable for the adaptation of cross-disciplinary pedagogies and can be a vehicle for the inclusion of different disciplinary perspectives.

The disadvantages can be summarized as follows:

– The programs and courses often lack links between climate change and heritage.

– Most courses do not have an interdisciplinary perspective.

– Innovative pedagogical approaches are rarely used in programs and courses.

– In most of the courses examined, heritage issues are secondary. They also fail to redefine heritage as a means to address the climate crisis.

Evaluating the potential of existing courses on the subject, updating their content and methods to address deficiencies and designing new courses are important to raise awareness of the use of cultural heritage as an asset to mitigate the impacts of climate change. Capacity building for climate action will be achieved by improving the knowledge and skills of future professionals.

List of Authors

Sibel ACAR
Department of Architecture
TOBB University of Economics
and Technology
Ankara
Turkey

Albena ANTONOVA
Faculty of Mathematics
and Informatics
Sofia University "St Kliment
Ohdridski"
Bulgaria

Lina BEKKER
Department of the Built Environment
Eindhoven University of Technology
(TU/e)
The Netherlands

Boyan BONTCHEV
Faculty of Mathematics
and Informatics
Sofia University "St Kliment
Ohdridski"
Bulgaria

Scott BREMER
Study of the Sciences
and the Humanities
University of Bergen
Norway

Jelle BUIJS
Faculty of Social Sciences
Leiden University
The Netherlands

Nihan BULUT
Protection and Repair of Cultural
Heritage Department
Faculty of Architecture
Izmir Institute of Technology
Turkey

Irene G. CURULLI
Department of the Built Environment
Eindhoven University of Technology
(TU/e)
The Netherlands

Yavor DANKOV
Faculty of Mathematics
and Informatics
Sofia University "St. Kliment
Ohdridski"
Bulgaria

Andjela DANKOVA
University of Architecture, Civil
Engineering and Geodesy (UACEG)
Sofia
Bulgaria

Luciano DE BONIS
Department of Biosciences
and Territory
Università Degli Studi Del Molise
(UNIMOL)
Campobasso
Italy

Yakim DEYKOV
Department of Drawing
and Modelling
University of Architecture, Civil
Engineering and Geodesy (UACEG)
Sofia
Bulgaria

Pelin GÜROL-ÖNGÖREN
Department of Architecture
TOBB University of Economics
and Technology
Ankara
Turkey

Paloma GUZMAN
Norwegian Institute for Cultural
Heritage Research (NIKU)
Oslo
Norway

Deniz IKIZ KAYA
Department of the Built Environment
Eindhoven University of Technology
(TU/e)
The Netherlands

Maya ISHIZAWA ESCUDERO
Independent heritage consultant
Bonn
Germany

Rohit JIGYASU
ICCROM
Rome
Italy

Eugene JO
ICCROM
Rome
Italy

Anna KAISER
Center for Cultural
Property Protection
University for Continuing
Education Krems
Austria

Arghavan KHAEFI
Department of the Built Environment
Eindhoven University of Technology
(TU/e)
The Netherlands

Irem KÜÇÜK
Department of Architecture
Gazi University
Ankara
Turkey

Lazaros MAVROMATIDIS
ICube UMR 7357 Laboratory
National Institute of
Applied Sciences
Strasbourg
France

Rossella NOCERA
Department of Biosciences
and Territory
Università Degli Studi Del Molise
(UNIMOL)
Campobasso
Italy

Giovanni OTTAVIANO
Department of Biosciences
and Territory
Università Degli Studi Del Molise
(UNIMOL)
Campobasso
Italy

Asli ÖZBEK
Department of Architecture
TOBB University of Economics
and Technology
Ankara
Turkey

Giovanni PARISANI
Department of Biosciences
and Territory
Università Degli Studi Del Molise
(UNIMOL)
Campobasso
Italy

Susan ROSS
Indigenous and Canadian
Studies Department
Carleton University
Ottawa
Canada

Eva STARK
Department of the Built Environment
Eindhoven University of Technology
(TU/e)
The Netherlands

Valentina TERZIEVA
Institute of Information and
Communication Technologies
Bulgarian Academy of Sciences
Sofia
Bulgaria

Maria Gabriella TROVATO
Department of Landscape
Architecture
Faculty of Landscape and Society
Norwegian University of
Life Sciences
Ås
Norway

Dessislava VASSILEVA
Scientific Research Department
Sofia University "St Kliment
Ohridski"
Bulgaria

Burçin YILMAZ
Department of the Built Environment
Eindhoven University of Technology
(TU/e)
The Netherlands

Index

D

E, F

G, H

I

L

V, W

Other titles from

ISTE

in

Science, Society and New Technologies

2023

CARDON Alain, EL HAMI Abdelkhalak
Fundamental Generation Systems: Computer Science and Artificial Consciousness, the Informational Field of Generation of the Universe, the Sixth Sense of Living Beings
(Digital Sciences Set – Volume 4)

ELAMÉ Esoh
The Sustainable City in Africa Facing the Challenge of Liquid Sanitation
(Territory Development Set – Volume 2)

JURCZENKO Emmanuel
Climate Investing: New Strategies and Implementation Challenges

PARET Dominique, CRÉGO Pierre, SOLÈRE Pauline
Smart Patches: Biosensors, Graphene and Intra-Body Communications

VENTRE Daniel, LOISEAU Hugo
Cybercrime during the SARS-CoV-2 Pandemic (2019-2022): Evolutions, Adaptations, Consequences
(Cybersecurity Set – Volume 3)

2022

AIT HADDOU Hassan, TOUBANOS Dimitri, VILLIEN Philippe
Ecological Transition in Education and Research

CARDON Alain
Information Organization of The Universe and Living Things: Generation of Space, Quantum and Molecular Elements, Coactive Generation of Living Organisms and Multiagent Model
(Digital Science Set – Volume 3)

CAULI Marie, FAVIER Laurence, JEANNAS Jean-Yves
Digital Dictionary

DAVERNE-BAILLY Carole, WITTORSKI Richard
Research Methodology in Education and Training: Postures, Practices and Forms
(Education Set – Volume 12)

ELAMÉ Esoh
Sustainable Intercultural Urbanism at the Service of the African City of Tomorrow
(Territory Development Set – Volume 1)

FLEURET Sébastien
A Back and Forth Between Tourism and Health: From Medical Tourism to Global Health
(Tourism and Mobility Systems Set – Volume 5)

KAMPELIS Nikos, KOLOKOTSA Denia
Smart Zero-energy Buildings and Communities for Smart Grids
(Engineering, Energy and Architecture Set – Volume 9)

2021

BARDIOT Clarisse
Performing Arts and Digital Humanities: From Traces to Data
(Traces Set – Volume 5)

BENSRHAIR Abdelaziz, BAPIN Thierry
From AI to Autonomous and Connected Vehicles: Advanced Driver-Assistance Systems (ADAS)
(Digital Science Set – Volume 2)

BÉRANGER Jérôme
Societal Responsibility of Artificial Intelligence: Towards an Ethical and Eco-responsible AI
Technological Prospects and Social Applications Set – Volume 4)

CORDELIER Benoit, GALIBERT Oliver
Digital Health Communications
Technological Prospective and Social Applications Set – Volume 5)

DOUAY Nicolas, MINJA Michael
Urban Planning for Transitions

GALINON-MÉLÉNEC Béatrice
The Trace Odyssey 1: A Journey Beyond Appearances
(Traces Set – Volume 4)

HENRY Antoine
Platform and Collective Intelligence: Digital Ecosystem of Organizations

KOLOKOTSA Denia, KAMPELIS Nikos
Smart Buildings, Smart Communities and Demand Response
(Engineering, Energy and Architecture Set – Volume 8)

LE LAY Stéphane, SAVIGNAC Emmanuelle, LÉNEL Pierre, FRANCES Jean
The Gamification of Society
(Research, Innovative Theories and Methods in SSH Set – Volume 2)

RADI Bouchaïb, EL HAMI Abdelkhalak
Optimizations and Programming: Linear, Non-linear, Dynamic, Stochastic and Applications with Matlab
(Digital Science Set – Volume 1)

2020

ALAKTIF Jamila, CALLENS Stéphane
Migration and Climate Change: From the Emergence of Human Cultures to Contemporary Management in Organizations

BARNOUIN Jacques
The World's Construction Mechanism: Trajectories, Imbalances and the Future of Societies
(Interdisciplinarity between Biological Sciences and Social Sciences Set – Volume 4)

ÇAĞLAR Nur, CURULLI Irene G., SIPAHIOĞLU Işıl Ruhi, MAVROMATIDIS Lazaros
Thresholds in Architectural Education
(Engineering, Energy and Architecture Set – Volume 7)

DUBOIS Michel J.F.
Humans in the Making: In the Beginning was Technique
(Social Interdisciplinarity Set – Volume 4)

ETCHEVERRIA Olivier
The Restaurant, A Geographical Approach: From Invention to Gourmet Tourist Destinations
(Tourism and Mobility Systems Set – Volume 3)

GREFE Gwenaëlle, PEYRAT-GUILLARD Dominique
Shapes of Tourism Employment: HRM in the Worlds of Hotels and Air Transport
(Tourism and Mobility Systems Set – Volume 4)

JEANNERET Yves
The Trace Factory
(Traces Set – Volume 3)

KATSAFADOS Petros, MAVROMATIDIS Elias, SPYROU Christos
Numerical Weather Prediction and Data Assimilation
(Engineering, Energy and Architecture Set – Volume 6)

MARTI Caroline
Cultural Mediations of Brands: Unadvertization and Quest for Authority
(Communication Approaches to Commercial Mediation Set – Volume 1)

MAVROMATIDIS Lazaros E.
Climatic Heterotopias as Spaces of Inclusion: Sew Up the Urban Fabric
(Research in Architectural Education Set – Volume 1)

MOURATIDOU Eleni
Re-presentation Policies of the Fashion Industry: Discourse, Apparatus and Power
(Communication Approaches to Commercial Mediation Set – Volume 2)

SCHMITT Daniel, THÉBAULT Marine, BURCZYKOWSKI Ludovic
Image Beyond the Screen: Projection Mapping

VIOLIER Philippe, with the collaboration of TAUNAY Benjamin
The Tourist Places of the World
(Tourism and Mobility Systems Set – Volume 2)

2019

BRIANÇON Muriel
The Meaning of Otherness in Education: Stakes, Forms, Process, Thoughts and Transfers
(Education Set – Volume 3)

DESCHAMPS Jacqueline
Mediation: A Concept for Information and Communication Sciences
(Concepts to Conceive 21st Century Society Set – Volume 1)

DOUSSET Laurent, PARK Sejin, GUILLE-ESCURET Georges
Kinship, Ecology and History: Renewal of Conjunctures
(Interdisciplinarity between Biological Sciences and Social Sciences Set – Volume 3)

DUPONT Olivier
Power
(Concepts to Conceive 21st Century Society Set – Volume 2)

FERRARATO Coline
Prospective Philosophy of Software: A Simondonian Study

GUAAYBESS Tourya
The Media in Arab Countries: From Development Theories to Cooperation Policies

HAGÈGE Hélène
Education for Responsibility
(Education Set – Volume 4)

LARDELLIER Pascal
The Ritual Institution of Society
(Traces Set – Volume 2)

LARROCHE Valérie
The Dispositif
(Concepts to Conceive 21st Century Society Set – Volume 3)

LATERRASSE Jean
Transport and Town Planning: The City in Search of Sustainable Development

LELEU-MERVIEL Sylvie, SCHMITT Daniel, USEILLE Philippe
From UXD to LivXD: Living eXperience Design

LENOIR Virgil Cristian
Ethically Structured Processes
(Innovation and Responsibility Set – Volume 4)

LOPEZ Fanny, PELLEGRINO Margot, COUTARD Olivier
Local Energy Autonomy: Spaces, Scales, Politics
(Urban Engineering Set – Volume 1)

METZGER Jean-Paul
Discourse: A Concept for Information and Communication Sciences
(Concepts to Conceive 21st Century Society Set – Volume 4)

MICHA Irini, VAIOU Dina
Alternative Takes to the City
(Engineering, Energy and Architecture Set – Volume 5)

PÉLISSIER Chrysta
Learner Support in Online Learning Environments

PIETTE Albert
Theoretical Anthropology or How to Observe a Human Being
(Research, Innovative Theories and Methods in SSH Set – Volume 1)

PIRIOU Jérôme
The Tourist Region: A Co-Construction of Tourism Stakeholders
(Tourism and Mobility Systems Set – Volume 1)

PUMAIN Denise
Geographical Modeling: Cities and Territories
(Modeling Methodologies in Social Sciences Set – Volume 2)

WALDECK Roger
Methods and Interdisciplinarity
(Modeling Methodologies in Social Sciences Set – Volume 1)

2018

BARTHES Angela, CHAMPOLLION Pierre, ALPE Yves
Evolutions of the Complex Relationship Between Education and Territories
(Education Set – Volume 1)

BÉRANGER Jérôme
The Algorithmic Code of Ethics: Ethics at the Bedside of the Digital Revolution
(Technological Prospects and Social Applications Set – Volume 2)

DUGUÉ Bernard
Time, Emergences and Communications
(Engineering, Energy and Architecture Set – Volume 4)

GEORGANTOPOULOU Christina G., GEORGANTOPOULOS George A.
Fluid Mechanics in Channel, Pipe and Aerodynamic Design Geometries 1
(Engineering, Energy and Architecture Set – Volume 2)

GEORGANTOPOULOU Christina G., GEORGANTOPOULOS George A.
Fluid Mechanics in Channel, Pipe and Aerodynamic Design Geometries 2
(Engineering, Energy and Architecture Set – Volume 3)

GUILLE-ESCURET Georges
Social Structures and Natural Systems: Is a Scientific Assemblage Workable?
(Social Interdisciplinarity Set – Volume 2)

LARINI Michel, BARTHES Angela
Quantitative and Statistical Data in Education: From Data Collection to Data Processing
(Education Set – Volume 2)

LELEU-MERVIEL Sylvie
Informational Tracking
(Traces Set – Volume 1)

SALGUES Bruno
Society 5.0: Industry of the Future, Technologies, Methods and Tools
(Technological Prospects and Social Applications Set – Volume 1)

TRESTINI Marc
Modeling of Next Generation Digital Learning Environments: Complex Systems Theory

2017

ANICHINI Giulia, CARRARO Flavia, GESLIN Philippe, GUILLE-ESCURET Georges
Technicity vs Scientificity – Complementarities and Rivalries
(Interdisciplinarity between Biological Sciences and Social Sciences Set – Volume 2)

DUGUÉ Bernard
Information and the World Stage – From Philosophy to Science, the World of Forms and Communications
(Engineering, Energy and Architecture Set – Volume 1)

GESLIN Philippe
Inside Anthropotechnology – User and Culture Centered Experience
(Social Interdisciplinarity Set – Volume 1)

GORIA Stéphane
Methods and Tools for Creative Competitive Intelligence

KEMBELLEC Gérald, BROUDOUS Evelyne
Reading and Writing Knowledge in Scientific Communities: Digital Humanities and Knowledge Construction

MAESSCHALCK Marc
Reflexive Governance for Research and Innovative Knowledge (Responsible Research and Innovation Set - Volume 6)

PARK Sejin, GUILLE-ESCURET Georges
Sociobiology vs Socioecology: Consequences of an Unraveling Debate (Interdisciplinarity between Biological Sciences and Social Sciences Set – Volume 1)

PELLÉ Sophie
Business, Innovation and Responsibility (Responsible Research and Innovation Set – Volume 7)

2016

ANDRÉ Michel, SAMARAS Zissis
Energy and Environment (Research for Innovative Transports Set – Volume 1)

BLANQUART Corinne, CLAUSEN Uwe, JACOB Bernard
Towards Innovative Freight and Logistics (Research for Innovative Transports Set – Volume 2)

BRONNER Gérald
Belief and Misbelief Asymmetry on the Internet

COHEN Simon, YANNIS George
Traffic Management (Research for Innovative Transports Set – Volume 3)

EL FALLAH SEGHROUCHNI Amal, ISHIKAWA Fuyuki, HÉRAULT Laurent, TOKUDA Hideyuki
Enablers for Smart Cities

GIANNI Robert
Responsibility and Freedom (Responsible Research and Innovation Set – Volume 2)

GRUNWALD Armin
The Hermeneutic Side of Responsible Research and Innovation
(Responsible Research and Innovation Set – Volume 5)

LAGRAÑA Fernando
E-mail and Behavioral Changes: Uses and Misuses of Electronic Communications

LENOIR Virgil Cristian
Ethical Efficiency: Responsibility and Contingency
(Responsible Research and Innovation Set – Volume 1)

MAESSCHALCK Marc
Reflexive Governance for Research and Innovative Knowledge
(Responsible Research and Innovation Set – Volume 6)

PELLÉ Sophie, REBER Bernard
From Ethical Review to Responsible Research and Innovation
(Responsible Research and Innovation Set – Volume 3)

REBER Bernard
Precautionary Principle, Pluralism and Deliberation: Sciences and Ethics
(Responsible Research and Innovation Set – Volume 4)

TORRENTI Jean-Michel, LA TORRE Francesca
Materials and Infrastructures 1
(Research for Innovative Transports Set – Volume 5A

TORRENTI Jean-Michel, LA TORRE Francesca
Materials and Infrastructures 2
(Research for Innovative Transports Set – Volume 5B)

VENTRE Daniel
Information Warfare – 2nd edition

YANNIS George, COHEN Simon
Traffic Safety
(Research for Innovative Transports Set – Volume 4)

www.ingramcontent.com/pod-product-compliance
Ingram Content Group UK Ltd.
Pitfield, Milton Keynes, MK11 3LW, UK
UKHW020250290926
15470UKWH00002B/850

9 781786 30903